MW00480189

ROMANIA'S ABANDONED CHILDREN

ROMANIA'S ABANDONED CHILDREN

Deprivation, Brain Development, and the Struggle for Recovery

Charles A. Nelson
Nathan A. Fox
Charles H. Zeanah

HARVARD UNIVERSITY PRESS
Cambridge, Massachusetts, and London, England
2014

Copyright © 2014 by the President and Fellows of Harvard College
All rights reserved
Printed in the United States of America

Library of Congress Cataloging-in-Publication Data
Nelson, Charles A. (Charles Alexander)
   Romania's abandoned children : deprivation, brain development, and the struggle for recovery / Charles A. Nelson, Nathan A. Fox, Charles H. Zeanah.
      pages   cm
   Includes bibliographical references and index.
   ISBN 978-0-674-72470-9 (alk. paper)
   1. Abandoned children--Romania--Psychology.   2. Abandoned children--Deinstitutionalization--Romania.   3. Deprivation (Psychology).
I. Fox, Nathan A.   II. Zeanah, Charles H.   III. Title.

   HV887.R6N45 2014
   362.7309498--dc23   2013017625

To Gwen and Colin; to Betsy, Rebecca, and David; and to
Paula, Emily, Katy, and Melanie . . . for everything

Contents

Preface

This book is the culmination of more than a decade of research on currently and previously institutionalized children, along with children from the community who were never institutionalized. Although the story takes place in Romania, the implications for children around the world are considerable. UNICEF estimates that as many as 70 to 100 million children worldwide are orphans, with 8 million living in institutions. (UNICEF defines an orphan as a child who has lost at least one parent.) The estimated number of children living in institutions is surely low, given that many poor countries relying on institutional care do not keep reliable records.

This number will most likely grow in coming years, for two reasons. First, wars and illnesses like AIDS will leave more children without parents. Second, the number of children being adopted internationally, at least by U.S. families, has dropped considerably (from a high of about 20,000 in the early 2000s to approximately 12,000 in 2010 to about 10,000 in 2012). In the face of such a large and complex public health challenge, the U.S. government has been exploring ways to implement evidenced-based strategies among children living outside of family care.

Understanding what happens to children who experience profound early deprivation is essential: it will improve our scientific understanding of the role of experience in brain development, and it will guide governments on how best to care for parentless children. Finally, it will provide much-needed information both to child-protection systems, which bear responsibility for children who are neglected or abused by their parents, and to families who take such children into their homes.

ROMANIA'S
ABANDONED
CHILDREN

Chapter 1

The Beginning of a Journey

Let's begin with a bold premise: that understanding the human brain holds the key to understanding all of human behavior, which in turn may unlock the mysteries surrounding many of the ills that have challenged societies for millennia. But suppose that our ability to understand the adult brain will never be possible unless we first understand brain *development*—that is, how the two-celled zygote, the product of one sperm and one egg, morphs first into a simple neural tube (which forms just a few weeks after conception) and then into the complex, three-pound organ that in little more than two years from conception propels an infant from a nonverbal being to one who squeals, "No!" while being chased by his mother, who wants to get him into the bath.

Although indirectly this book is about brain development, it is more accurately about how experience—or rather, the lack of experience—impacts the course of brain development and, therefore, child development. It is the story of children abandoned by their parents and reared in state-run institutions. With only modest reframing, it is really about what happens to brains and people when certain fundamental expectations are not met—expectations, for example, that infants and young children will be exposed to sights and sounds (to stimulate hearing and vision), that an adult will comfort them when they need comforting, that adults will talk to them (to teach them language and acknowledge their presence), and that adults will provide the basic care necessary given our inability to take care of ourselves in our youngest years. These needs and responses happen so routinely in typically developing children that we take them for granted. But this story is about children being raised in an environment where responses do not necessarily match needs.

Our understanding of brain development has increased exponentially over the past two decades. These gains can be attributed to advances made in the neurosciences. Animal models have shed light on everything from the genes involved in building a brain to the molecules involved in building a neural circuit. Similarly, our ability to image the living, human brain has expanded by leaps and bounds, to the point that we can now noninvasively peer inside the brain of a newborn during his sleep to examine the brain's anatomy as well as its electrical and metabolic activity. And, though much remains to be discovered about the details of brain development, we are now in a position to make several assertions with great certainty. First, we know that brains are built over time, beginning a few weeks after conception and continuing through mid-to-late adolescence/early adulthood. Thus, despite the misconception that resurfaces periodically, brain development does *not* end at age three.

Second, we know that much as an architect supplies a blueprint for a house, genes supply the initial blueprint for the development of the brain. This genetic plan determines the basic properties of nerve cells as well as the basic rules for interconnecting nerve cells within and across circuits. In this manner, the genetic blueprint sets up the template for brain architecture.

Third, once genes have provided the framework for all subsequent brain development, experience begins to play a critical role in fine-tuning the brain. Although this influence begins prenatally (the effects of prenatal alcohol exposure or maternal stress are just two examples of the effects of experience during the prenatal period), fine-tuning as a result of experience becomes enormously important once a child is born, and it continues through adolescence. The neuroscientist William Greenough proposed two mechanisms by which experience weaves its way into the structure of the brain: experience-expectant development and experience-dependent development.[1]

Experience-expectant development refers to the process by which experiences that occur during a narrow window of time early in development have a significant influence on subsequent development. As a rule, these experiences are common to all members of the species and include, for example, access to patterned light (to facilitate visual development), access to faces and speech (to facilitate social communication and language), and access to appropriate caregiving. In contrast, experience-dependent development refers to changes that occur in the brain throughout the lifespan and are unique to each individual. Learning and memory are examples of this form of development.

Finally, we know that the timing of experience plays an essential role in many aspects of brain development, particularly (though not exclusively) those elements that occur after birth.[2] The principle here, then—a theme carried forward throughout this book—is that of a *sensitive period*. Specifically, for many brain functions, genes confer basic structure, but because of the limited number of genes (current estimates place this figure at about 20,000), it is advantageous for our adaptation to allow experience to affect gene expression to regulate many elements of brain development, from simple to complex circuits, from sensory cortices to association cortices. Thus genes code for the basics and experience does the fine-tuning. As we will see in the following pages, when this principle is violated, brain development can be undermined, leading to profound alterations in behavioral development.

Politics and Policy

Neuroscientists have known for several decades that brains are built over time, that some domains of development are more dependent on experience than others, and that within those domains the timing of experience is critical to healthy development. This information has not always received the attention it deserves, particularly among policymakers and advocates charged with safeguarding vulnerable children (for example, those seeking to protect children in the child-protection system or children who have been abandoned or orphaned in countries that do not facilitate permanent placements).

On April 17, 1997, President Bill Clinton and First Lady Hillary Rodham Clinton hosted "The White House Conference on Early Childhood Development and Learning: What New Research on the Brain Tells Us about Our Youngest Children." This conference spurred tremendous interest in the importance of early brain development. However, perhaps because there was only one neuroscientist at the conference or because the media took a simplistic approach to the information, or because the conference coincided with the publication of some recent research on learning, the core messages of brain development were largely misconstrued. For example, a paper published about this time demonstrated that college-age students who listened to Mozart for a few minutes showed short-term improvement in their spatial ability.[3] Soon thereafter, the governors of Georgia and Michigan began sending Baby Mozart™ CDs home with new babies, and a multi-million–dollar industry was born, purportedly designed to facilitate infant brain development (think Baby Einstein™). Suddenly, the public was inundated with news articles about so-called critical periods and the "fact" that brain development is "all over" by the age of three. Indeed, in one tongue-in-cheek editorial in the *New York Times,* a new mother lamented that now that her child was three years old, the toddler's fate was sealed.[4]

The media coverage that followed the White House conference led to a spirited public discussion about brain development. Some scientists complained that the work discussed at the conference was actually not new at all, but that some of this information had been known for many years.[5] Others took the information as a call to arms, a challenge to change public policy to ensure that all children had a healthy start in life. Ironically, child development experts had been arguing this same point for decades, but once images of the brain began to appear in the popular press, the need for intervention was given greater weight.

With this background, in 1997 the John D. and Catherine T. MacArthur Foundation expressed interest in supporting interdisciplinary efforts to shed light on the key issues pertaining to early experience and brain development. In February 1998, the foundation formally launched a research network entitled "Early Experience and Brain Development," which Charles A. Nelson directed and in which Charles Zeanah and Nathan Fox were core members.

The study this book details came about as a direct result of this network.[6]

Guiding Principles

The opening position of the group included the following two principles:

1. Experience is the product of an ongoing, reciprocal interaction between the environment and the brain.
2. Individuality is the product of both personal experience and biological inheritance.

Experience

Experience has been defined traditionally by the properties of the environment in which an individual lives. For example, experience might be characterized as exposure to a particular method of teaching or immersion in comfortable, stimulating surroundings. Science tells us, however, that experience is not simply a function of the environment but also the result of a complex, two-way interaction between that environment and the developing brain.

Within this context, the impact of any given experience can vary enormously under identical environmental conditions, depending on the history, maturation, and state of the individual's brain. For example, listening to a lecture spoken in Chinese will be a completely different experience for a person who understands Chinese compared with one who does not, for a three-year-old child compared with an adult, and for an individual who is interested in the subject compared with one who is indifferent. This principle, which is so obvious when considered in the context of complex experiences, applies equally to simple ones. Even an apparently simple physical experience (for example, an infant being gently tossed in the air by her father, as babies frequently are) may vary widely depending on the background and state of the individual involved (most adults probably do not like the feeling of being tossed in the air, especially, for example, if it occurs during a particularly turbulent plane ride).

The relative maturity of the brain also has a great impact on experience. Different areas of the brain mature at different rates, with sensory-processing areas maturing earlier than areas supporting complex cognition. A young child who is exposed to information before his or her brain is capable of processing that information will not have the same experience as an adolescent who has more advanced capability. As the brain matures and changes with experience, it is influenced by more complex cognitive interpretations of the environment. Thus, as an individual's brain changes, particularly during early development, the same physical environment can result in very different experiences. Language acquisition is a good example. We know that the complexity of children's language, including their vocabulary, depends heavily on the language to which they are exposed. However, using complex sentence structure and "big" words will have much less of an impact on a six-month-old than it will on a three-year-old, owing to different levels of brain maturation at these two ages—the brain of the six-month-old simply cannot make use of this more advanced input.

Finally, certain properties of the brain differ dramatically across individuals and within individuals over time. Therefore, because experience is defined as the interaction of the brain with the environment, a scientific description of an experience must include a description of the context in which experience occurs, the developmental stage of the infant or child, including the maturity of the brain, and a description of the specific experience to which the individual is exposed.

There is a valuable lesson here: infants and children are not passive recipients of information, and experiences do not simply happen *to* them, no matter how young they are or how passive they may look to adult eyes. Rather, what children bring to the experience matters a great deal. What kinds of things are we talking about, exactly? A short list includes children's developmental and biological history, their cultural niche, the status and integrity of the brain when the experience occurs, and gradually, how they come to interpret the experience. Two children can grow up under seemingly identical conditions and yet have very different developmental outcomes.

A study of complex reciprocal interactions, such as those described above, requires a longitudinal design and measurement of multiple domains. If properly conducted, it points us to differences in outcomes for young children who begin life in similar conditions of risk. In the case of infants who all were abandoned and lived in contexts of severe psychosocial deprivation, these were our explicit goals in the Bucharest Early Intervention Project (BEIP).

Individuality

The brain develops according to a complex array of genetically programmed influences. These include both molecular and electrical signals that arise spontaneously in growing neural networks. Together these signals establish neural pathways and patterns of connections that are remarkably precise and that make it possible for animals to carry out discrete behaviors beginning immediately after birth. They also underlie instinctive behaviors that may appear much later in life, often associated with emotional responses, foraging, sex, and social behavior.

Genes specify the properties of neurons and neural connections to different degrees in different pathways and at different levels of processing. The extent of genetic determination reflects the degree to which the information processed at a particular connection is predictable from one generation to the next. Because many aspects of an individual's world are not predictable, the circuitry of the brain relies on experience to customize connections to serve his or her needs. Experience shapes these neural connections and interactions, sometimes powerfully, but always within the constraints imposed by genetics.

The impact of experience on the brain is not constant throughout life. Early experience often exerts a particularly strong influence in shaping the functional properties of the immature brain. Many neural connections pass through a period during development when the capacity for experience-driven modification is greater than it is in adulthood. Language skills, emotional responses, and social behavior, as well as basic sensory and motor capacities, are shaped powerfully and, in many cases, permanently, during these sensitive periods. Thus individual capabilities reflect the combined influences of both evolutionary learning and personal experience.

In the BEIP we have seen domains that are closely tied to sensitive periods (for example, language, attachment) and others less so (executive functions, psychopathology). We have also seen domains in which the efficacy of the intervention is not bounded by time; in other words, development is not constrained by early experience. Thus we are confronted with a dilemma faced by many neuroscientists and psychologists: the interrelationship of dose, timing, duration, and specificity of context; in other words, *how much of* an experience is critical to facilitate typical development, the *timing* of that experience, the *duration* of that experience, and what domain of expertise or behavior it affects. These four factors influence both the effects of early institutionalization and the efficacy of our foster care intervention.

Having established some common assumptions on which its work would be based, the "Early Experience and Brain Development" network began to build its research agenda. The central question evident to the members of the group was, How do different experiences impact the developing brain and, downstream, an individual's behavior? At first, animal studies dominated the work of the group, simply because it is easier to manipulate experience and examine the effects on brain and brain development in animals than in humans. Here we conducted work with rodents, song birds, barn owls, and importantly, rhesus monkeys. Indeed, as we describe in a later section of this chapter, our research on monkeys, led by Dr. Judy Cameron, set the stage for our work with institutionalized children in Romania.

Early Psychosocial Deprivation and Human Development

Orphaned or abandoned children have for centuries been raised in institutions. In the mid-twentieth century, a series of important studies showed that children raised in institutional settings suffered from a variety of developmental sequelae, ranging from stunted growth to intellectual impairment to emotional disturbances, including what Rene Spitz referred to as "hospitalism" (characterized by miasma, listlessness, low arousal, lack of social responsiveness, and lack of affective expression).[7] These effects are reviewed in detail in

Chapter 6. Although children raised in institutions are clearly deprived of many experiences typical of other children, most scientists assumed that lack of consistent, sensitive caregiving—so-called psychosocial deprivation—was at the top of the list. We have known for many years that nurturing relationships with adults, beginning at birth, provide an important foundation upon which all subsequent relationships are built. The stable, loving, and secure bonds an infant forms with her caregivers confer on the child a range of competencies, including a strong sense of self, positive social skills, successful intimate relationships later in life, and a sophisticated understanding of emotions.[8]

Research from the neurosciences (mostly done with rodents but some with nonhuman primates) has recently demonstrated that such early positive relationships help to build and strengthen brain architecture. For example, work from Michael Meaney's lab, at McGill University, has shown that infant rats born to and cared for by mothers who engage in a high degree of licking and grooming (the rodent manifestation of good maternal care) grow up to be less anxious, handle stress better, and possess greater cognitive abilities than rats whose mothers lick and groom less.[9] At a molecular level, the density of a nerve receptor (glucocorticoid receptor) in the part of the brain involved in memory (hippocampus) differs between the rat pups receiving good maternal care and those who received low levels of good care (which appears to account for differences in the ability to respond to stress).[10] Intriguingly, female rats born and raised by high-licking and -grooming mothers tend to grow up to be high-licking and -grooming mothers themselves.[11] This intergenerational transmission is not genetic in the conventional sense —the infant rats do not "inherit" high-licking and -grooming behavior—rather, it is *epigenetic,* meaning that maternal experience impacts the way DNA in the infant brain is expressed, which in turn alters brain structure and function.[12]

Scientific discovery is often serendipitous—opportunities present themselves in unexpected ways. So-called accidents of nature represent such an opportunity, and one became available to us as the network organized plans for research. We were given the chance to conduct a study on young children in Romania that had certain parallels to one of our signature projects with rhesus monkeys.

First, a brief history of the context of our study in Romania (for more details, see Chapters 3 and 6). In 1989, the Romanian dictator Nicolae Ceauşescu was overthrown in a coup and subsequently executed. Suddenly Romania, a Communist country for the previous forty-four years, and a place that few in the West knew much about, was opened to the world. ABC News broke the story that enormous numbers of children were being raised in state-run institutions—eventually it became clear that the number was as many as 170,000 children—and many of them were poorly cared for.[13] As the Western press reported extensively on the plight of such children, many were quickly adopted by parents in Western Europe and North America. Sadly, many families were ill prepared for the extent of the children's disabilities. A paper in the *Journal of the American Medical Association* reported on the condition of children living in the Romanian institutions, concluding vividly: "There is a desperate need for an immediate international response to the current orphanage crisis in Romania. Thousands of young lives are currently being jeopardized and potentially lost. These children are not unsalvageable and labeling them as such has done them a grave injustice."[14]

Dana Johnson, a pediatrician at the University of Minnesota and a colleague of Charles Nelson's, had created the first international adoption clinic in the United States in 1986, and so had considerable experience with children who had been raised in institutions. Nelson and Johnson frequently discussed the problems that Johnson was seeing in children adopted from institutions, including many from Romania. Johnson was invited to attend a meeting of the MacArthur Network, at which he described the profound developmental delays and abnormal social-emotional behavior that plagued these children. Several members of the network recognized an unsettling parallel between these children and the maternally deprived monkeys in Harry Harlow's famous experiments of the 1950s and 1960s.[15] Harlow raised monkeys in socially deprived environments, sometimes with peers and sometimes not, but always without their mothers. He observed in his monkeys something eerily similar to what Johnson described in previously institutionalized children—abnormal attachment behaviors with emotional constriction, unusual stereotypies (purposeless, repetitive motor movements such as rocking or twirling), and significantly compromised intellectual function.

These findings also converged with observations made by Judy Cameron, a neuroscientist in the network, who had begun a network-sponsored project about the effects of maternal deprivation on *Rhesus macaques*. In this project monkeys were cage-reared together in large social groups. In most laboratory-based studies of monkeys, infants are separated from their mothers at six months of age. In the experiment Cameron designed, groups of infants were separated at one week, one month, three months, or six months of age. Instead of removing infants from their mothers, as Harlow had done, Cameron kept the infants in their social groups and instead removed the mothers from the group. The infants were then left to be raised by the other monkeys.

Cameron and her group made two startling observations: first, the monkeys whose mothers were removed at one week or one month appeared to be very psychologically handicapped. Second, these two groups differed dramatically in precisely how the experience of maternal removal affected them. Monkeys separated at one week showed profound disinterest in and concern for other monkeys, preferring to spend their time alone, often rocking or engaging in other stereotypies. In contrast, the monkeys separated at one month had an intense need to be with other monkeys and displayed heightened anxiety, clinging indiscriminately to whichever other monkey was available. In contrast to the animals separated at one week, which were profoundly anxious when around other monkeys, the animals separated at four weeks were anxious when *not* around other monkeys. Their clinging behavior extended well beyond the period for which clinging to other animals was normative.

As the group pondered these findings, we concluded that the difference in the timing of separation affected the consequences of the social bond disruption. In the case of the animals separated at one month, the relationship between the infant and its mother was disrupted; in the case of the monkeys separated at one week, this bond was never established to begin with.

Although we are not comparing directly the behaviors of infant monkeys with human children, an eerily similar profile was described by Johnson in children he had witnessed in institutions and in previously institutionalized children now adopted into families in the United States. Many of these children, despite being adopted into loving homes, displayed a range of attachment behaviors that were very atypical for children their age. These behaviors were more extreme, but also similar to, the behaviors Charles Zeanah, a child psychiatrist, had seen among severely neglected children in the United States. For example, some severely neglected children displayed distress without seeking or responding to comfort, extreme emotional constriction, a tendency to approach and engage with strangers, and violation of social boundaries. Could the network launch a study of children who had spent their early lives being raised in large institutions, without a primary caregiver?

The Plight of Children Who Begin Life in an Institution

In the late 1990s, as we were preparing to launch the BEIP, an unprecedented number of previously institutionalized children from Romania were welcomed into homes in the United States, Canada, and the United Kingdom. Investigations were launched to study these children. This work complemented the studies completed in the mid-twentieth century, reporting that the longer children were reared in institutions, the more altered their development and the more impaired their behavior.[16] Unlike previous work, however, this new group of studies involved state-of-the-art measures and more sophisticated research designs aimed at determining the effects of the timing of certain experiences on development (see Chapter 6).

We were following this important work closely and were impressed by the magnitude of the effects on young children of being raised in institutions, as well as by the degree of recovery that seemed possible for some, though not all. Still, we knew that children who were adopted were often selected by prospective adoptive parents (or selected *for* prospective parents by adoption agencies or orphanage directors), and so were often the healthiest or most attractive or most engaging children among their peers. It was hard to know how much this potential for bias affected reported findings, but it remained as a limitation nonetheless.

After Dana Johnson spoke at a network meeting, the idea began to germinate that we might be able to conduct a rigorous scientific study examining the effects of early institutionalization on brain and behavioral development. Indeed, Dana suggested we make contact with Cristian Tabacaru, then minister of child protection in Bucharest. At the time Tabacaru was seeking ways to persuade members of the government to pass legislation that would close down institutions throughout the country, and he thought scientific evidence might be more persuasive than arguments on ethical or moral grounds. To evaluate the possibility of conducting such a study in Bucharest, Charles Zeanah made an initial trip to Bucharest in December 1998. Upon his return the network spent the better part of 1.5 years considering different scientific designs as well as the ethical issues underlying any project that might eventually be launched. Out of these discussions was born the Bucharest Early Intervention Project. At the outset the project had two primary goals: 1. to examine the effects of institutionalization on brain and behavioral development; and 2. to examine whether the effects of early institutionalization could be reversed by placing children in families. The study was uniquely positioned to adopt a rigorous experimental design—randomized controlled trial—that would permit us to test three specific hypotheses: First, the development of children raised in families would be enhanced compared with that of children raised in institutions. Second, the longer children remained in institutions, the more compromised their development was likely to be. Third, the *age at placement* in our foster care intervention families might prove to be more important in long-term outcome than the *duration of time* in these families (Table 1.1 lays

Table 1.1   Long-term outcomes of different kinds of care

	Hypothetical long-term findings	Implications
Baseline	IG = NIG	Adverse early experiences do not matter; individual differences are likely intrinsic. As a result, regardless of rearing environment, all groups will perform similarly on the BEIP battery.
	IG < NIG	Children who are placed in institutions have compromised development, either because of prenatal experiences that are associated with institutional placement or because early postnatal experiences of deprivation compromise developing systems.
Follow-up	CAUG = FCG = NIG	Early deficits, if they occur, are effaced by subsequent development, and as with the "baseline" predictions, rearing environments do not contribute to outcome.
	CAUG = FCG < NIG	Early experiences determine outcomes; remediation through environmental enhancement is not possible.
	CAUG < FCG = NIG	Early experiences of deprivation are completely remediable through environmental enhancement.
	CAUG < FCG < NIG	Adverse early experiences have effects that are only somewhat remediable through environmental enhancement, and may depend on timing of intervention.

Note: CAUG, care-as-usual (prolonged institutional care) group; FCG, foster care intervention group; IG, institutionalized group (children in institutions prior to randomization); NIG, never-institutionalized community comparison group. See Chapter 2 for details about the experimental design of BEIP.

out some of the possible outcomes and their interpretation). If our hypotheses were confirmed, the findings would have major policy implications, particularly in Romania but potentially beyond. For example, perhaps governments would consider alternative interventions to institutional rearing. With millions of children worldwide orphaned, abandoned, or maltreated, many as a result of AIDS in Africa and the conflicts in the Middle East and Africa, the best way to care for orphans remains an urgent concern.

Through all our planning, however, we were well aware that even the best scientific hypotheses were not always confirmed. In the case of BEIP, we had to consider the possibility that *if* the children who were placed in institutions were indeed somehow different from those who were not abandoned, then taking such children out of institutions and placing them into families might not prove efficacious, for the simple reason that they began life already too disadvantaged/compromised. Here, it is important to note that when we first started the project, we heard from some in Romania that our study was unnecessary, because the state did a better job of raising children than did families; and children who were abandoned to institutions were by default "defective" children. (In Chapter 3, we discuss in more detail the origins and implications of this belief.) We knew that if government officials uniformly believed these premises, they would certainly conclude that there was nothing wrong with institutional rearing, and that raising children en masse was a very prudent solution to the problem of orphaned or abandoned children. Thus the stakes were high when we implemented this project.

Chapter 2

Study Design and Launch

Undertaking a large-scale intervention project outside the United States, with a vulnerable population (institutionalized children), was a daunting task. Our desire to assess a range of competencies in these children, to do this in a central location in Bucharest, and to include measurements of brain activity all required a specially designed laboratory, special equipment, and a staff well trained to work with this population. In addition, our intervention required a professional staff of social workers who would work with the foster families we had recruited. These skilled social workers would help us not only to identify qualified foster care parents but also to monitor and support these families as the study progressed. Finally, we needed to cultivate a network of Romanian colleagues who could serve as liaisons between us and government officials. Given that none of us spoke a word of Romanian, we would be heavily dependent on our staff to help us form alliances.

First Steps

We knew from the outset that we needed to identify an on-the-ground leader of this project, someone to hire and oversee the staff and to represent our project at the highest levels of the Romanian government and to the various nongovernmental organizations (NGOs) with whom we eventually partnered. Sebastian Koga came to fill this role admirably, far exceeding our expectations. Sebastian was a twenty-two-year-old third-year medical student at an international medical school in Cambridge, England, in late 1999. At the time Sebastian was considering taking a leave from school, so when he heard that a group of U.S. investigators was implementing a project in Bucharest and looking for a project manager, he expressed interest in the position.

Sebastian was a native Romanian who had emigrated to the United States with his mother and sister after the revolution of 1989, when he was twelve years old. His father had had difficulties with the Communist authorities and had escaped from Romania before the fall of Ceaușescu, and the entire family had reunited in the United States after 1989.

Sebastian was hired, and in September 2000 he moved to Bucharest to begin work on the BEIP. Having a bilingual and bicultural project manager proved key to the success of our enterprise. Sebastian was involved in every aspect of the project, from overseeing construction of the laboratory, to negotiating with Romanian authorities, hiring staff, repairing the EEG (electroencephalogram) equipment, recruiting foster parents, and many other tasks. His gift was attention to detail coupled with a full grasp of the big picture; it also didn't hurt that he was highly intelligent. Sebastian stayed with the project through June 2004, after which he completed medical school in the United States and is now a neurosurgeon.

The Launch

Given the enormity of our undertaking, we decided it would be wise to conduct a feasibility study, which we launched in fall 2000, once Sebastian was in place and our initial staff was hired. The goal of this phase of the research was not only to decide on our methods and procedures but also to train the staff in their use and to pilot these measures with children from the community who had never been institutionalized and were not targeted for the study itself (in the end, this involved approximately fifty children).

The first research assistant we hired was Anca Radulescu in the fall of 2000. She had been working for two years as a psychologist at St. Catherine's Placement Center, the largest institution for young children in Romania, at the time we hired her. Originally a nurse, Anca had worked with cancer patients and became interested in the psychological aspects of their care. She returned to school to become a psychologist, which in Romania required a bachelor's degree. After a year of looking for work, she was hired to work at St. Catherine's in 1998, shortly after major legislation reformed the child-protection system (see Chapter 3). At the time she was hired, St. Catherine's had become the site of the newly constructed BEIP laboratory. Having Anca involved was invaluable from the outset. She understood the inner workings of St. Catherine's, was familiar with policies and procedures, knew key personnel, and had a good working relationship with child-protection authorities.

Soon after, we hired Carmen Calancea as a second research assistant, and, as the feasibility phase proceeded successfully, we hired the first social workers early in 2001, Alina Rosu and Veronica Ivascanu.

One unique aspect of our study was the measurement of brain activity and brain responses to stimuli among infants living in the institutions (prior to the study onset) and then during the follow-up stages. In order to accomplish these measurements we had to build a laboratory for electrophysiological assessment. While the use of EEG equipment in hospitals was common, no one to our knowledge was completing research studies acquiring EEG and computing event-related potentials to stimuli, and certainly no one was doing this with infants and young children. Nathan Fox had a number of years of experience designing laboratory equipment for the acquisition of these signals with infants, and he and his then post-doctoral fellow, Peter Marshall, set out to establish a lab in St. Catherine's. Peter and Nathan traveled to Romania with the hardware necessary for this set-up (amplifiers, monitors, computers) and wired the laboratory for EEG acquisition. Getting the hardware through customs and then setting it up in the space given to us by Child Protection were adventures by themselves, but by the time we were ready for the baseline assessments, the EEG equipment was fully functional. In addition, we sent two of our research assistants, Anca Radalescu and Nadia Radu, to the University of Maryland, where they were given a crash course in EEG acquisition methods. These preparations helped make it possible for us to record brain activity over the course of the study.

Figure 2.1 illustrates our general time line. By March 2001, we had built the laboratory, hired the first staff members, and completed enough piloting of measures to conclude that the project

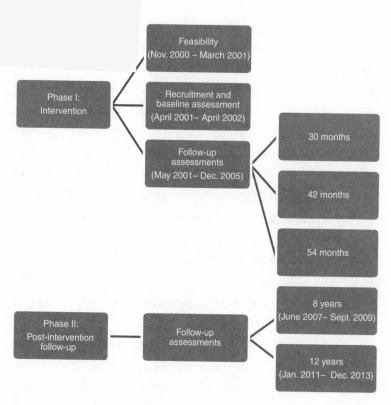

Figure 2.1. Time line of the BEIP.

was indeed feasible. On April 1, 2001, the formal intervention was launched. We immediately added two new research assistants, Adela Apetroia and Nicoleta Corlan, followed by Iuliana Dobre and a third social worker, Amelia Graceanu.

The Institutions

Similar to Paris, Bucharest is divided into numbered sectors, or districts. Each sector had its own institutions, and we recruited children from five of the six sectors. There was some variation across institutions—for example, some were larger than others (St. Catherine's, which is where the project was based, was the largest, in 1989 housing 850 children younger than three years old). In addition, there was also some variation within and across institutions in physical makeup; for example, within St. Catherine's, some wards had been decorated with colorful paint and mobiles, whereas others were painted stark white, with no decorations whatsoever. The units, each headed by a physician, were characterized by a sense of instrumental efficiency but little sensitivity or individualized care. Mealtimes were silent affairs, with one caregiver often feeding young toddlers out of a large common bowl, spooning in the "meal" to each child in turn. Bath times were often difficult, conducted with assembly-line efficiency but little sensitivity and often characterized by young children crying loudly before, during, and after their turn at being scrubbed.

Across most of the institutions from which we recruited, the ratio of children to caregivers (generally between 12:1 and 15:1) was the same, as was the fact that all caregivers were public employees with little education and little training in child development. Given these conditions, they could hardly be expected to invest much in the care of their young charges.

Participants

Institutionalized Group

At the time we launched the project, there were more than 100,000 institutionalized children in Romania, approximately 4,200 of them in Bucharest. Although UNICEF had reported that the vast majority of children who were abandoned to institutions in Romania were there because of poverty, because their mothers were young and unwed, or because the children were "different" from other children at birth (for example, they were premature or had a birth defect), similar circumstances exist throughout the world, and not all such children are abandoned by their mothers.

It is widely accepted that the social-engineering policies of Ceauşescu contributed to the high rate of child abandonment in Romania. Under his leadership, families were reassured that the state would step in to raise their children. According to Romanian law at the time, as long as the parents visited their child in the institution at least once every six months, they retained some legal rights over that child. In this respect Romania's abandoned-child problem differs in some ways from that of other countries. However, because we do not have information about *why* mothers abandoned their children, we are unable to address the interesting question as to how the families who abandoned children differed from families of similar ethnic and socioeconomic status who did not abandon their children. Our random assignment to foster care or institutional care addresses the issue of sample bias *out of* institutions but does not address the potential sample bias of who *enters* an institution (a bias for which there is no ethical workaround).

Four of Bucharest's six sectors had a *leagan* (Romanian word for "cradle") or institution where infants and young children were housed, and Sector 1 had two. We recruited from all six institutions to ensure adequate representation of institutionalized children (that is, to avoid the possibility that children from one institution might somehow differ from those at another). Our inclusion criteria for these children were that they had to be under the age of thirty months and had to have spent at least half of their lives living in an institution. Before being enrolled in the study, all children were screened by Dana Johnson, a pediatrician, and Mary Jo Spencer, a pediatric nurse practitioner, with a pediatric exam that included neurological tests, growth measurements, and assessment of any physical abnormalities. We excluded children with known genetic syndromes such as Down syndrome, obvious signs of fetal alcohol syndrome, and microcephaly (more than 2.5 standard deviations, or *SD,* below the mean for occipitofrontal circumference). Given the relatively small sample of children, we were mindful that recruiting those with very serious limitations in their capacity to recover could potentially skew our results, making them difficult to interpret. So though we tried to include as many children as possible, those with serious limitations had to be excluded.

We originally screened a total of 187 children from the six institutions but excluded 51 on the basis of the criteria listed above. The remaining 136 children were in reasonably good health and met our inclusion and exclusion criteria. Some problems, of course, cannot be identified easily in infancy. For example, at entry to the study none of the children was thought to have cerebral palsy (CP), but several years later, we identified one child, in fact, with CP. Mild forms of cerebral palsy can be difficult to detect very early in life, and this child had been evaluated as a young infant.

Although we carefully screened children for overt signs of genetic/chromosomal and neurological abnormalities, we remained concerned about more subtle factors that might impact development, in particular, prematurity. Premature and low–birth weight babies carry their own set of risk factors. Unfortunately, gestational ages of the children were not consistently reliable, and we had medical records for only 112 children. These records indicated that children were born between thirty and forty-two weeks (forty weeks is a full-term delivery). Birth weight was a more reliable indicator, but it was available only for 122 children. Birth weights ranged from 900 g, or just under 2 lbs., to 4,500 g, or about 10 lbs. ($M = 2,790$ g, $SD = 609$ g), which differed from that of the children in the community–control group ($M = 3,333$ g, $SD = 459$ g).

Of the final sample of 136 children, 52 percent were of Romanian ethnicity, 35 percent were Roma, and the remaining 13 percent were mixed or could not be classified. All children at the time of the baseline assessment had spent between 51 and 100 percent of their lives in the institution, with 52 percent having spent their entire lives institutionalized.

Our survey of infants living in the institutions in Bucharest found that the youngest were around six months of age at our initial screening; at the time, many abandoned newborns were maintained in maternity hospitals until around six months of age, when they were transferred to *leagane* (the plural of "leagan"). Thus the average age of our sample at the baseline assessment was twenty-two months, with a range from six to thirty-one months of age. This sample characteristic would have a significant effect on our study. Recall that one of our primary questions was whether there were sensitive periods in development during which our intervention would be more efficacious in remediating the problems that we found in young children growing up in institutions. Because all the children in our study were between the ages of six and thirty-one months when we began, we would have no way of detecting sensitive periods below the age of six months. On the other hand, the period of six to thirty-one months is a time of rapid brain development, so we were hopeful that we could find evidence of such sensitive periods within this age range in our assessment measures.

After the baseline assessment, infants were randomized to one of the two arms of the study (either intervention or continued institutionalization). Children had varied in the age at which they had entered the institution (some immediately after birth, some after a period of six months), and so the duration of their experience with deprivation and the age at which the intervention began varied across the sample. Subsequently, this meant that at follow-ups, each conducted at a particular age (for example, thirty, forty-two, fifty-four months), the children varied in the duration of intervention they had received.

> **Groups in the BEIP**
>
> CAUG   care-as-usual group
> FCG   foster care group
> IG   institutionalized group
> NIG   never-institutionalized group (community sample)

Community Comparison Group

Although we planned to compare children who were placed in foster care (the foster care group, or FCG) with those who received care as usual (the care-as-usual group, or CAUG), it was vital that we also recruit a typically developing Romanian comparison sample. Most of the measures we used had never before been used in Romania, and we needed to know if typically developing children in Romania performed similarly or differently from typically developing children in the United States, on whom most of these measures were normed. Fortunately, the Romanian community children and their American counterparts appeared quite similar on nearly all measures. As a result, we then had at our disposal an in-country comparison sample of typically developing Romanian children against whom to compare both children in foster care and those receiving care as usual.

With this in mind, we recruited seventy-two children who had never been institutionalized and who had been born at the same maternity hospitals where the foster care group and the care-as-usual group had been born. The parents of these children were approached by personnel from the Institute of Maternal and Child Health (IOMC), with whom we had established a partnership. During routine pediatric clinic visits, IOMC workers invited parents to participate in the study. These children were similar in age to the institutionalized children and roughly equivalent in the distribution of boys and girls. They also underwent a similar screening. All community children (referred to throughout as the never-institutionalized group, or NIG) fell within 2 *SD* of the mean for physical growth (weight, length, occipitofrontal circumference). There were forty-two males and thirty-one females in the NIG; 92 percent were Romanian, 6 percent were Roma, one child was Spanish, and one child was Turkish.

Randomization

Following comprehensive baseline assessments, children in the institutionalized group were randomly assigned either to the foster care intervention (discussed in greater detail in Chapter 5) or to care as usual (that is, continued institutional care). Random assignment was determined by placing identification numbers into a hat (one for each child) and then drawing out one piece of paper at a time. The one exception to randomization involved siblings, whom we kept together for ethical reasons, writing two numbers on a single piece of paper before placing it in the hat.

Sixty-eight children were assigned to foster care (FCG) and sixty-eight children were assigned to care as usual (CAUG). We compared children in the two groups before placement and found no differences in gender, ethnicity, age, birth weight, or percentage of life spent in an institution.

We established some guidelines with government authorities at the start of this randomized trial. Among them was the agreement that we would not interfere with changes or transitions that might occur with children in either group (foster care or care as usual) over the course of time. For example, the authorities might decide to remove a child from an institution and reunite him with his biological family or place him in some alternative living situation. The same might be true for a child we placed in BEIP foster care. Throughout this book, unless otherwise noted, all our data have been analyzed using a so-called *intent-to-treat* design, meaning that regardless of where a child is currently living, we analyzed that child's data as though the child was in the originally assigned group. Figure 2.2 illustrates the original group assignments as well as the changes in group assignments through age eight.

Study Design

Assessing children at baseline before randomization assured us that the intervention and control groups were comparable on a number of important measures. This in turn increased our confidence that any differences in outcome would reflect true intervention effects. Moreover, randomization prior to intervention made it very likely

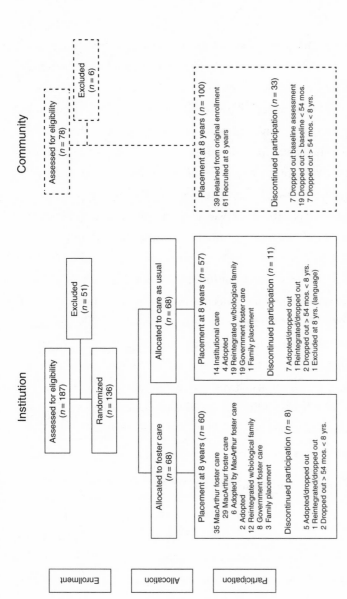

Figure 2.2. Group assignment at age eight. As the figure demonstrates, over the course of the study children's group assignments tended to change. Since the responsibility for these changes fell to the Romanian child-protection authorities and not to the BEIP investigators, in nearly all cases we adopted an "intent to treat" design in our data analyses, thereby avoiding the potential for sample bias.

Community

Assessed for eligibility
(*n* = 78)

Excluded
(*n* = 6)

Placement at 8 years (*n* = 100)

39 Retained from original enrollment
61 Recruited at 8 years

Discontinued participation (*n* = 33)

7 Dropped out baseline assessment
19 Dropped out > baseline < 54 mos.
7 Dropped out > 54 mos. < 8 yrs.

Institution

Assessed for eligibility
(*n* = 187)

Excluded
(*n* = 51)

Randomized
(*n* = 136)

Allocated to care as usual
(*n* = 68)

Allocated to foster care
(*n* = 68)

Placement at 8 years (*n* = 57)

14 Institutional care
4 Adopted
19 Reintegrated w/biological family
19 Government foster care
1 Family placement

Discontinued participation (*n* = 11)

7 Adopted/dropped out
1 Reintegrated/dropped out
2 Dropped out > 54 mos. < 8 yrs.
1 Excluded at 8 yrs. (language)

Placement at 8 years (*n* = 60)

35 MacArthur foster care
29 MacArthur foster care
6 Adopted by MacArthur foster care
2 Adopted
12 Reintegrated w/biological family
8 Government foster care
3 Family placement

Discontinued participation (*n* = 8)

5 Adopted/dropped out
1 Reintegrated/dropped out
2 Dropped out > 54 mos. < 8 yrs.

Enrollment

Allocation

Participation

that prenatal and other risk factors (for example, iron deficiency, lead exposure, drug exposure), which could not be determined with certainty, would be evenly distributed across the two groups.

One additional aspect of our design should be reiterated. Because children ranged in age from six to thirty-one months at baseline, and because they were then followed up at distinct age points (thirty, forty-two, fifty-four, and ninety-six months), at each assessment point we could evaluate the effects of timing of placement on their outcomes. Timing of placement into foster care confounds age at which the placement occurred and amount of time spent in foster care. For example, a child randomized at twenty-six months had received four months of foster care intervention at thirty months. In contrast, a child placed at sixteen months had received fourteen months of foster placement at age thirty months (see Figure 2.3). If the latter child has a better outcome at thirty months, we cannot say whether the difference is due to the child's having received the intervention at a younger age (a timing effect) or to his having received ten additional months of foster care (a dose or length-of-time-in-care effect).

The Institutional Environment

Institutions around the world differ enormously, and the same can be said for the six institutions from which we recruited children in Bucharest. Still, these six institutions shared many features, including the background and education/training of the staff, a regimented daily schedule, a high ratio of children to caregivers, and a management structure led by medical personnel.

Our laboratory was based at St. Catherine's Placement Center (referred to in the Communist era as Bucharest #1), in Sector 1. At

Length of intervention

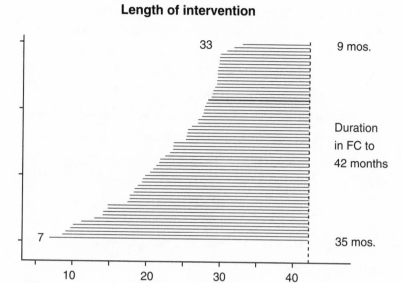

Figure 2.3. The length of intervention varied as a function of age at placement into foster care.

the time of Ceauşescu's trial and execution in 1989, approximately 850 children were living at St. Catherine's (Bucharest no. 1). By the time our study began, there were approximately 450. These children ranged in age from about six months to six years in fall 2000; some had profound disabilities (spina bifida, neurosyphilis, hydrocephalus) and others had no obvious abnormalities.

Although infants and toddlers shared slightly different daily schedules, all children were woken at 6:30 A.M., were put down for a nap from 1 to 3 P.M., and were put to bed at 8 P.M. As a rule children were fed meals in small groups seated in high chairs or around small tables. Washing and changing children were typically scheduled before and after meals. The youngest children were given ninety minutes of "stimulation" each morning, whereas children older than three attended educational activities or the equivalent of kindergarten. In good weather, outdoor playtime was scheduled before lunch and before dinner. Figures 2.4–2.6 provide glimpses into the children's rooms and play areas.

Beginning in the early 1990s, humanitarian aid supplied most institutions with toys and playgrounds; however, in our observations, caregivers generally monitored children rather than actually played with them.

Despite a highly regimented schedule, the children had limited opportunities to interact socially with caregivers. They had long stretches of unstructured playtime, in which caregivers mostly chatted with one another, while large numbers of children wandered aimlessly. Over the course of a day, a child might have contact with

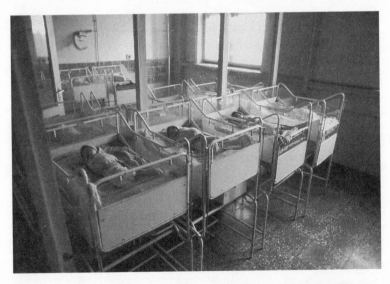

Figure 2.4. Infants in cribs in an institution in Bucharest. (Photograph by Michael Carroll)

Figure 2.5. The playground at St. Catherine's from the vantage point of the original BEIP lab. The swimming pool in the center of the yard was used for both recreation and bathing. (Photograph by Charles A. Nelson)

multiple adults, but without much consistency or opportunity for sustained interaction. In a given week, a child may have come in contact with seventeen different caregivers, three housekeepers, four nurses, two educators, a physician, a psychologist, and a physical therapist. Our analyses, based on ratings of videotaped naturalistic interactions in institutions and homes, indicated substantially lower quality of care provided by institutional caregivers than by parents of community children.[1]

Some of our assessments required us to identify the child's primary caregiver, and in the case of the care-as-usual group, this was determined through interviews and observations. When the staff was unable to identify a favorite caregiver, we included the person

Figure 2.6. A typical toddler room at St. Catherine's. (Photograph by Charles A. Nelson)

who had spent the most time with the child and knew the child best.

From baseline through forty-two months of age, each assessment consisted of up to fourteen tests and measurements (depending on the age of the child) and was usually divided into three visits at the laboratory and one home or institutional observation. Additionally, physical growth measures of all children in the institutional and foster care groups were obtained monthly through age forty-two months and then again at eight years; for the never-institutionalized group, growth measurements were taken at each assessment point (thirty and forty-two months). At fifty-four months another assessment was implemented, focused mostly on IQ and mental health status. Finally, at eight and twelve years of age, another comprehensive battery of tests was administered (across three sessions), consisting of approximately twenty-five measures. As this book goes to press, we have just completed a twelve-year follow-up (and a sixteen-year follow-up is planned), so we focus here on data collected through eight years of age.

BEIP, it is important to note, is unique among studies of institutionalized children in that it is grounded in neuroscience. Indeed, we not only examined sensitive periods in development but also employed a range of state-of-the art measures of brain structure and functioning, including electroencephalograms (EEGs) and magnetic resonance imaging tests (MRIs), and eventually more molecular approaches, including the study of genetics and epigenetics.

Conclusion

BEIP is the first-ever randomized controlled trial of foster care as an intervention for early institutionalization. Its strengths as a study include randomization, a focus on brain-imaging methods and molecular biology, the use of a comprehensive battery of state-of-the-art psychological measures and methods, and an in-country comparison group. The study also has some limitations. First, we have very little information as to why mothers abandoned their children, nor do we have information about a child's prenatal history. These data would have informed us about the preinstitutional history of the children in the study. Second, children did not usually arrive at institutions from maternity hospitals until they were six months of age. Maternity wards served essentially as a holding station for children prior to placement in an institution, but for all intents and purposes, these wards were run exactly the same as institutions.

Nevertheless, we do not have infants between birth and six months in our sample. As a result, any claims we make about sensitive periods must be qualified by this fact and by the distribution of ages at randomization. Third, though we were successfully able to randomly assign children *to* foster care, we obviously did not randomly assign children *to* institutions. Children who were abandoned at birth and placed in institutions may not have been the same as children born into comparable families (for example, in level of poverty) who were not abandoned to institutions. We must remember that most poor families did not abandon their children. Fourth, for ethical reasons (see Chapter 4), over the course of the study children moved from one setting to another (as when an institutionalized child was reunited with his or her biological family), complicating some of our group comparisons. For example, three children were adopted within Romania almost immediately after the baseline assessment, one in care as usual and two in the foster care group. Over time, many children were moved from institutional settings into families, either their own biological families or government foster families, which did not exist in 2001 when we began. Data for all these children were analyzed on the basis of original group assignment rather than current placement. Therefore, results likely underestimate effects of foster care placement on children's development. A final limitation is that in attempting to juxtapose the effects of early experience against those of subsequent life experience, we were limited in some analyses and conclusions by the small size of our sample.

Chapter 3

The History of Child Institutionalization in Romania

> Orphans should be placed under the care of public guardians.
> Men should have a fear of the loneliness of orphans and of
> the souls of their departed parents. A man should love the
> unfortunate orphan of whom he is guardian as if he were his
> own child. He should be as careful and as diligent in the
> management of the orphan's property as of his own or even
> more careful still.
>
> Plato (Laws, 927)

In December 1989, the Romanian Communist dictator Nicolai
Ceauşescu was overthrown and executed. In the wake of the up-
heaval, Western media, which had come to Romania to cover the
revolution, discovered a vast network of institutions housing thousands of infants, children, and adolescents, many in deplorable conditions. American audiences learned of the situation on October 5,
1990, when the ABC news magazine *20/20* ran a special report on
these institutions.[1] Images of neglected, deprived, and frightened
children, some tied to metal cribs or cots, horrified the world. Romanians, too, were shocked by the appalling conditions of the insti-

tutions, and the newly formed government began immediate efforts to correct the situation. At the same time, NGOs from many countries arrived to help improve the facilities, and many families from the West came to Romania hoping to rescue these children through adoption.

In this context the Bucharest Early Intervention Project was conceived and launched. The project began in late 1999, just as major reforms in child protection in Romania were finally legislated. One of the key figures advocating these reforms, Cristian Tabacaru, Romania's secretary of state for child protection and head of the Romanian Adoption Agency, was also instrumental in facilitating the BEIP. To understand the situation of Romanian orphans and what led to the project, we must first understand how Romania came to create a national system of institutions for infants and young children, the societal conditions under which families abandoned their children to be raised by the state, and the legacy of these policies for Romania in the decades after the end of Communism.[2]

Institutions as a Societal Intervention

Orphanages existed in what is now Romania as early as the first century. During the Middle Ages, bishops in that area recommended that orphaned children be placed in monasteries. Some orphanages practiced what was called "binding-out," in which children, as soon as they were old enough, were given as apprentices to households. Orphanages were often viewed as safe havens for lost children and were supported by the upper classes with charitable donations and by the religious establishments of different localities.

They were seen not only as benign but also as facilitating adaptive growth and citizenship in young children.[3]

Institutions for abandoned and orphaned children are described in Europe in the Middle Ages and were well known in the Renaissance. The Ospedale degli Innocenti in Florence, for example, was built for the care of abandoned children and admitted its first infant in February of 1445.[4] In many parts of the developed world, however, these institutions were uncommon until the eighteenth century. Most abandoned children who could not be informally fostered by relatives or neighbors typically lived as street children or were placed in almshouses for the poor along with adults. In North America, Ursuline nuns founded the first orphanage in the United States in New Orleans, Louisiana, in 1729 after the slaying of adult settlers near Natchez, Mississippi.[5] In Romania, there are references to orphanages in Bucharest as early as 1798, and in other parts of the country during the nineteenth century.[6]

Orphanages became more common in both the United States and Europe in the nineteenth century, owing to the growing numbers of children made parentless by disease and war and, in the United States, by expanded urbanization and immigration. Such changes increased the number of poor families.[7] Moreover, the move from rural to urban areas led to a loss of support from the extended family and, consequently, the need for childcare, as parents had to work.[8] According to one report, in the eighteenth century 10 to 40 percent of children in European cities were abandoned.[9] Among the abandoned were four children fathered by Jean Jacques Rousseau (1781/1953). Historians debate exactly why Rousseau did this (his mental instability, his belief that he would make a bad father), but it is ironic that the philosopher who wrote one of the more influential books on the nature of the child and education *(Emile on Education)* would abandon his children in this way. Indeed, in *Emile* he writes, "Your first duty is to be human. Love childhood. Look with friendly eyes on its games, its pleasures, its amiable dispositions."[10]

As industrialization began to flourish at the end of the nineteenth century, orphanages appealed to a complicated mix of desires to reach out to the less fortunate, to instill wholesome values in the young, and perhaps to remove "undesirables" from the public.[11]

It was not until the mid-twentieth century that American investigators such as Rene Spitz and William Goldfarb published studies indicating that institutional care was harmful to children's development, although the pediatrician Henry Dwight Chapin had called for an end to institutions near the turn of the century on account of their high mortality rates.[12] The British child psychiatrist and psychoanalyst John Bowlby published a monograph for the newly created World Health Organization (WHO) in which he argued that care by a mother figure was not only preferred but essential for children's mental health.[13] Foster care as an alternative to institutional care developed rapidly in the second half of the twentieth century in both the United States and the United Kingdom.[14] Although most orphanages in both countries had closed by the 1970s, they have never completely disappeared in the United States. These days, however, they house only about 0.5 percent of children who are in state custody, and these are intended to be short-term placements.[15] In Western Europe, foster care is common, although given knowledge about the research on the harmful effects of institutions it is surprising that thousands of children younger than three years of age are raised in institutions there.[16]

During the second half of the twentieth century, Romania became a part of the Soviet bloc and was effectively cut off from Western psychology research and child-protection practices which included recognizing the harmful effects of institutional rearing. The country was influenced by Soviet doctrine regarding communal child rearing that emphasized the importance of productivity and the subjugation of the self for the greater good of the state. Communal rearing allowed the state to maintain greater control over the population and also to separate productive and nonproductive members of the community. Isolation from Western research precluded Romanian academics from learning about new research in child development, particularly about the importance of parental care during the early years of life. This provided fertile ground for the disastrous ideological policies that followed and made possible an unprecedented promotion of government-sponsored child rearing in large institutions.[17]

Brief History of Romania

Located at the crossroads of East and West, Romania has been conquered and occupied by the Roman Empire, Asian Huns, the Ottoman Empire, and the Austro-Hungarian Empire, among others. Romania's people did not become a united and independent nation until the second half of the nineteenth century. Contemporary Romania was not consolidated until the twentieth century.

The region of Transylvania, in the northeast part of the country, on the Hungarian border, was ruled by Roman Catholic Austro-Hungarian kings from medieval times until 1921, when it became part of Greater Romania. The regions of Moldova and Wallachia were protectorates of Orthodox Russian tsars on and off since

1779, when Russia was expanding its territory toward the Balkans. These royal powers influenced the response of the church and the state to the problem of child abandonment.[18] Children were to be protected, particularly in times of political and military strife.

In the Second World War, the Romanian government joined with Germany in the fight against the Soviet Union. As the war progressed and the defeat of Germany became apparent, King Michael of Romania led a revolt against the fascist government in Bucharest and established a government that pursued peace with the Allied forces. At the conclusion of the war, Romania (by agreement with the other Allies) was occupied by Soviet forces, who remained there until 1947, removing the king and his government and establishing a Communist-led government, which remained in control for more than forty years. The country became part of the Soviet bloc, behind the Iron Curtain, and many of its social and medical structures followed Soviet designs. The Soviets forced Romania to send many natural resources and food products to the Soviet Union without compensation, thereby increasing Romania's poverty and its dependence on the Soviet Union.[19]

Ceauşescu's Rule

Following the death in 1965 of Gheorghiu-Dej, then premier of Romania, Nicolai Ceauşescu, a former cobbler who had worked his way up through the ranks of the Communist Party, assumed leadership of the country. In Ceauşescu's vision, Romania needed more workers to boost industrialization and increase Romania's wealth and power. He was determined to create a modern industrial economy that would supply goods to the entire Communist world.

Soon after coming to power, Ceauşescu became known in the

West as a maverick among Soviet bloc leaders, exhibiting independence from Soviet influence. He ended Romania's active participation in the Warsaw Pact (though Romania formally remained a member), and he condemned the Soviet invasion of Czechoslovakia in 1968. Eager to drive any possible wedge between Europe and the Soviets, President Richard Nixon visited with Ceaușescu in Bucharest in 1969, and Romania was granted most-favored-nation trading status by the United States in 1975.[20]

To fulfill his vision of making Romania a leader among Communist industrial states, Ceaușescu undertook a massive program of urbanization, moving people from their homes in the country to the cities to live in monolithic blocks of identical apartments in identical buildings so they could work in Romanian factories. Villages were razed to make way for new agro-industrial complexes. The worker, rather than the family, became the fundamental unit of society, and the state penetrated almost every aspect of family life.[21] Relocated to housing in many different areas, extended families were separated. Even spouses might be assigned to job sites that were far apart.[22]

At the time Ceaușescu assumed power, Romania had one of the highest rates of abortion and divorce in Eastern Europe and a declining birth rate.[23] To counteract these trends and generate human capital, Ceaușescu implemented a set of policies designed to increase the population. These policies began in the mid-1960s with the passage of a law (Decree 770/1966) that banned abortion, among other things.[24] A law passed in 1966 (Decree No. 779) restricted divorces to exceptional cases. For the next twenty-four years Ceaușescu attempted to force women to be "heroes of socialist labor" by having four or five children.[25] In addition to banning abortion and restricting divorce, the government set up what have been termed "pronatalist" incentives. These were provided from the state to working mothers. Money was paid to families to help them support their children. Cash allowances rose as family size increased. Families in cities received more than those living in the country-side.[26] The government rewarded women who bore many children, calling them "Heroine Mother"; women with seven, eight, or nine children were awarded the "Order of Maternal Glory"; the "Maternal Medal" went to those with five or six children.[27]

Ceaușescu tried a number of initiatives to raise the birth rate. Among them, women were penalized for *not* having children, and a "childlessness tax" was imposed on young couples. In addition, the legal age for marriage was lowered to fifteen years.[28] It is interesting how involved the government became in the reproductive rights and experiences of women, including raising the minimum age for abortion back up to forty-five and restricting abortion to women who had at least five children.[29]

Ceaușescu recruited gynecologists to staff a division of the Department of State Security that came to be known as the "Menstrual Police." They conducted interrogations and gynecological exams—taking women from their workplaces or schools to examine them. This was to ensure that pregnant women carried their babies to term and to coerce nonpregnant women into having children. Again, as part of the government's intruding into the reproductive lives of women, Ceaușescu had reported miscarriages investigated.[30] If women had not had children, or did not seem to be producing enough children fast enough, they were interrogated and threatened. More than a decade after the revolution, someone assisting our project, then thirty-two years old, painfully described the humiliating experience of being examined and questioned by the Menstrual Police while still in high school.

These policies were designed to weaken the family as a cohesive unit. Parents were told that the state could raise children better than they could.[31] The government employed medical and mental health professionals to legitimize their attempt to convince parents to relinquish custody of their children to large, government-run institutions.

The economic situation was severe in the 1980s, and Ceauşescu imposed severe restrictions, resulting in significant shortages for the population. Many children were abandoned or voluntarily left in institutions by their parents.[32] In many cases, placement of the child was intended to be a temporary measure until the child could be supported at home, but often parents left their children expecting that the state would raise them, educate them, and make them into productive members of society. Raising children at home was difficult as the extended family was destroyed. Women worked, and there was little childcare. By the 1980s, most young couples moved to cities while their parents remained in the countryside, so they could no longer depend on their extended families to help with childcare. Childcare services were unavailable, particularly for those from low-income families.[33]

Child Institutions in the Ceauşescu Era, 1965–1989

The government passed a law (3/1970) which established different types of institutions for both typically developing and handicapped children.[34]

Leagane were designated for all abandoned and orphaned children from birth to three years. Children who were abandoned at birth in maternity hospitals were transferred to a leagan within months after birth. Other children who were abandoned after being with their parents for varying amounts of time also were admitted to a leagan—as long as they had not yet turned three years of age. When children in a leagan reached the age of three, they were transferred to another state-run institution on the basis of their developmental status. Upon turning three, children were examined by a pediatrician and/or psychiatrist/psychologist and, based on results of the examination, were separated into two distinct groups at a "switching center." We interviewed a number of physicians about this exam, but we could not identify a systematic approach to this decision making process. Children who were judged to be physically and developmentally "normal" went to institutions known as *casi de copii* (children's homes). These were state-run residential institutions for preschool children (aged three to six). Children in these group homes attended public schools and were cared for by a rotating staff. From there, if all went well, they were transferred to casi di copii for school-aged children (seven years and up), where they remained until they were eighteen. Here too, the children were cared for by staff who worked rotating shifts, but the children continued to attend public schools. Approximately 60 percent of the children within the orphanage system lived in these facilities.[35]

At age three those children who were judged incapable of ever entering the workforce or who had special physical or mental needs were transferred to a *gradiniţa* (a residential special education institution for children deemed to have "curable" deficiencies) or to an *institutul neuropshichiatric* (a residential institution for preschool children with "incurable" deficiencies). When they turned seven, some entered a *camin spital* (a residential institution for school-age children deemed to have "incurable" deficiencies). The camin spital and institutul neuropshichiatric—institutions for the handicapped—were places where the most egregious deprivation occurred. Children there were grossly undernourished and mortality rates were high.[36]

The government also linked different types of institutions to different ministries. Greenwell describes these divisions: "The casi de copii were under the Ministry of Education; the camine spital were under the secretary of state for the handicapped; and the special education and vocational institutions were under the Ministry of Labor."[37]

Defectology: The Rationale

The policy of separating children into those who the state believed would be useful and productive and those who would not, like similar policies regarding the handicapped in other Eastern European counties, had its roots in the Soviet science of "defectology."[38] Developed in the 1920s, defectology defined disabilities as disease states in individuals that made them defective or abnormal. The emphasis was on intrinsic abnormalities—the environment was considered unimportant. Thus, if children were born defective, there was no point in providing interventions for them because they would never become productive. Detecting defects in children by age three or four would allow the state to segregate damaged children from productive children, who were more deserving of resources. Although morally repugnant and scientifically wrong, defectology underlay the system implemented by the Ceauşescu regime.

The science of defectology was developed at a time when deterministic models of development were in vogue throughout psychology, in the West as well as in the USSR. The belief was that individuals were born with certain characteristics and traits that were for the most part immutable. Studies refuting this approach at the time were rare. One exception was an intervention study by Harold Skeels in the 1930s that demonstrated that children deemed mentally retarded living in orphanages in the United States could in fact recover with appropriate intervention (removing them from the orphanages and placing them with women living in an institution).[39] Although Skeels's study was dismissed for thirty years in the United States because it defied commonly held beliefs about child development, such research eventually changed the attitude of U.S. psychologists toward the power of intervention and early experience. The tragedy for children in Eastern Europe is that this shift in approach to child development did not happen there for the next thirty to forty years. This underscores the potentially devastating effects of the scientific isolation that shrouded Romanian psychology after the Second World War.

Institutions for Typically Developing Children

The caregiving contexts of leagane varied with the size of the institution. The leagan at Bucharest #1 (previously and later named St. Catherine's), the largest leagan in Romania, housed more than 500 young children in 1998, but according to records on site, it had had 850 children a decade before that. Infants were kept in a small room with about 12 babies, and they spent a good part of their day lying in cribs with little stimulation or attention. They were fed regularly on a strict schedule, but they received extra holding or social interaction only if they were a caregiver's "favorite."

In a visit to Bucharest #1 in December 1998, we found a room with twelve cribs in which infants, ranging in age from six to eight months, were awake, lying quietly and passively on their backs. Although we could make eye contact with them, it took considerable effort to get them to smile at us—with several we never succeeded. The age of these infants placed them at the height of what Daniel Stern called the "most purely social time in development," and yet they were remarkably socially impaired.[40] These infants obviously had little or no opportunity for face-to-face interaction with a caregiver.

Caregiver-child ratios in the Ceauşescu era varied widely by location, but the ratios were appalling across the board. In some settings there might be one caregiver for twelve or fifteen infants, but in others it might be one caregiver for even more. Caregivers received no formal training and were generally assigned to one or more rooms depending upon the age of the children. Children were transferred to a toddler room once they could walk and were generally kept with their age mates.

Perhaps related to the large ratios was the detached manner in which caregivers typically interacted with children, even more than ten years after the fall of Ceauşescu. At outside playtime at St. Catherine's, for example, all forty-eight toddlers from two "units" might share the same play area and time. The three or four caregivers in charge mostly chatted with one another as the children wandered around the area, laid on blankets rocking, or perhaps interacted with one another. Indoor playtime during cold or inclement weather was even grimmer. Twelve or fifteen or even twenty toddlers might be in a large, mostly barren room supervised by one caregiver who either focused on her "favorite" or simply watched, intervening only occasionally. The children ignored their mates or fought briefly but fiercely for possession of the few available toys.

Other activities also lacked individual engagement. At mealtime at St. Catherine's, for example, a group of toddlers were seated around a small table in high chairs as a caregiver spooned out a bite to each child in turn from a large bowl. The process was mostly silent. Bath times, by contrast, were occasions for loud protest. These were handled with assembly-line efficiency and detachment. The children were lined up as one caregiver undressed them, another doused them with water and scrubbed them vigorously with soap, and a third dried them with a towel. Most of the children cried throughout this process. Several years later, when some of the children in our study were placed into foster homes, bath times were noted by foster parents to be especially difficult. Sometimes, it took weeks or longer for children to be bathed without distress.

Institutions for the Handicapped

The gradinițe, institutul neuropshichiatric, and camine spital were starker than leagane and casi de copii in every respect. Dana Johnson reported that children in these types of institutions were poorly cared for.[41] They were fed bread, fatty sausage, tripe gruel, and wormy apples. Facilities were often unheated. Medical care, educational services, and rehabilitation programs were nonexistent, and institutional sanitation and personal hygiene were ignored. In Videle, we witnessed children who squatted over plastic pots to defecate, and after completing their task, they simply pulled up their pajama pants without wiping or washing. The children did not look for assistance from caregivers, nor was any offered.

In one institution, 52 percent of children tested positive for HIV-1, and 60 percent tested positive for hepatitis B.[42] At a facility we visited in Bacau in 1998, 100 percent of the children were HIV positive according to the institution director. In addition, early reports by Western physicians who provided assistance at these facilities soon after the revolution documented malnutrition, growth retardation, lack of treatment for injuries, and overt physical and sexual abuse.[43]

In 1989, an epidemic of nosocomial HIV infection (that is, infection contracted during medical treatment) was discovered predominately among institutionalized children in Romania. Most likely they were infected through transfusions of unscreened blood and injections with improperly sterilized equipment. Plasma and whole blood transfusions were widely used among institutionalized children for a variety of ailments, including to "strengthen" them. This led to an epidemic of HIV and acute hepatitis B infections in these children.[44] The Communist government refused to recognize the epidemics, and in the period immediately before Ceaușescu's fall, there was a sharp increase in infant mortality attributed to causes such as "respiratory illness" or "endocrine and metabolic disorders." By 2000, 60 percent of Europe's pediatric HIV/AIDS cases were in Romania, and most of them occurred in infants and children living in institutions.[45] By this estimate, close to 10,000 children were infected. Once Ceaușescu was overthrown and the magnitude of the problem recognized, Romania responded with a significant effort to treat infected children and to prevent new cases of infection. The Romanian medical community partnered with U.S. government medical organizations, including the National Institutes of Health (NIH), the Centers for Disease Control (CDC), and USAID, to provide appropriate treatment of infected children. The programs have proven successful. By August 2003, 452 children were receiving antiretroviral (ARV) therapy monitored through outpatient clinics. Daily hospital admissions decreased to almost none, and childhood mortality sharply declined.[46]

The fact that vast numbers of children were institutionalized during this period raises a troubling question: How much did people know about the conditions in which children were being raised? Not all of the institutions were out in the countryside away from the eyes of the public. Some, such as St. Catherine's, were in the heart of Bucharest. Gabriella Koman, who headed the National Authority for Child Protection from 2000 to 2004, put it this way:

> I came into the child-protection system in 1992 without knowing very much about the system, because I think at that moment, no one is quite sure about what the crisis [meant], because all of us discovered, together with people from abroad, the shocking images from the previous orphanages from Ceaușescu's period without knowing [anything] about them . . . about the children placed there. This is not an excuse, of course, but anyhow, it's, for those coming to work in the system, it's an explanation for the motivation at that period. (G. Koman, interview, January 2011)

The child-protection system in the Ceaușescu era had little to do with protecting children and everything to do with protecting Romanian economic productivity. Resources for raising abandoned children were allocated with the goal of forming productive socialist citizens.[47] This was, after all, the real purpose of defectology—to separate those who would be productive from those who would not.

A 1992 study by Rosenberg conducted in a camin spital with children with severe deficiencies reported terrible conditions and a lack of medical care. The staff was small, and those who were there were untrained.[48] Even six years after this study was published, our own observations in facilities in Videle, Siret, and Bacau made it clear that despite considerable efforts there was still much work to be done.

Child Protection after Ceauşescu: 1989–2000

Immediate Responses

Following Ceauşescu's execution and the revelations of the tragic child-protection system he had created, Romanians tried to cope with three problems simultaneously: 1) tens of thousands of institutionalized children; 2) continuing high rates of child abandonment; and 3) the longer-term problem of developing a new approach to abandoned children. Over the next decade (1990–2000), there were a great many changes in the care and status of children living in institutions, as well as in the incidence of child abandonment and international adoption. For several years, abandonment and institutionalization in Romania actually increased because the collapse of the Communist regime exacerbated poverty and unemployment. International adoption was, at various times, included or excluded as a method of child protection.[49]

In December 1989, estimates of the total number of children living in institutions varied from 50,000 to 170,000.[50] Reports described conditions of significant deprivation. Infants spent all day in cribs and older children remained in huge rooms of ten to fifty children.[51] A report by an NGO at the time stated that "children suffered from inadequate food, shelter, clothing, medical care, lack of stimulation or education, and neglect. Disabled children suffered even grimmer conditions and treatment, with many malnourished and diseased."[52]

Most of the people we interviewed who worked during that time for child protection said they knew that institutions existed, but that they did not know about the huge number of children involved or about how terrible conditions were. For example, Gabriella Koman suggested that though a few doctors and relatives of those who worked in institutions must have known about the conditions in which the children were living, she believed that most Romanians were simply unaware of the "real situation."[53]

The overthrow of Ceaușescu and the establishment of a democratic government unfortunately did not result in an end to child abandonment. Infants and young children continued to be admitted to leagane, and the legal framework that promoted institutional care as a solution for families in difficulty remained unchanged. Throughout the decade after Ceaușescu's overthrow, the country continued to see a high rate of institutionalization.[54]

This is not to say that the Romanian government did not respond to the revelations of abuse and neglect among institutionalized children. The first response was to provide basic supplies, food, and medical and social services to the children. The government also welcomed aid from other countries and relief organizations.

NGOs from many Western countries established offices in Romania and began to pour resources into the country. Efforts were poorly coordinated and unevaluated for the most part, but the urgency of the revelations of children living in horrible conditions led to widespread support. Almost inevitably, Romania sought to have many of its institutionalized children adopted.

Conflicts among National Adoption, Foster Care, and International Adoption

To understand the debates about abandoned children in the decade after the overthrow of Ceaușescu, it is important to take into account a number of factors influencing this complex dynamic. In Romania, as elsewhere in Europe, kin adoption and fostering were prevalent, but adoption of nonkin was rare. It was common for grandparents, uncles, aunts, and other biological kin to adopt or foster an infant or child orphan. It was rare for families to adopt children who were not biologically related to them. These customs were not limited to Romania. In medieval society, bloodlines and biological relationships were paramount. Indeed, in both English Common Law and France's Napoleonic Code nonbiological adoption was either not permitted or made extremely difficult. Because of the restrictions on the adoption of abandoned infants by nonkin, the Church organized systems of foundling homes or institutions to house them. Across Europe, institutional care gained acceptance.[55] Thus it would have been unusual in Romania for families to adopt institutionalized children to whom they were not related. Add to this the fact that the infants and young children living in these institutions were often seen not simply as unfortunate but also as somehow defective (hence the reason for placement in institutions in the first place), and the situation becomes clear.

Romania had no established system of foster care. Although foster care was not illegal, and in some instances was even encouraged during the Ceauşescu era, it was rare for families to take in an institutionalized child. The administrative processes were cumbersome and did not encourage the practice. With the overthrow of Ceauşescu and the knowledge of conditions in institutions now broadcast around the world, many agencies and private individuals entered the country to facilitate international adoption. Here, too, however, the system was rife with problems.

Adoption of Abandoned Children

The law that governed what was done to abandoned infants (Law No. 3/1970) established care institutions for infants and children whose parents were unable to care for them. With the exposure of the terrible conditions in which children were being housed in institutions, Romania needed to find an alternative for the thousands of children who were being warehoused.

Seven months after the revolution, on July 31, 1990, the Romanian government passed a presidential decree legalizing and decentralizing international adoptions. The Law on Approval of Adoption (Law No. 11/1990) permitted international adoption and transferred the responsibility for adoptions from the president to the local courts. Courts, however, had no specific legal procedures to follow that would allow them to work with the then-nascent child-protection system. The result was that children were adopted in diverse ways. Foreign adoption agencies came to Romania and worked directly with directors of institutions. Child welfare NGOs worked with the courts in an attempt to provide some basic processes for international adoptions. Many families just came over and met directly with individual families privately, adopting from these biological parents by way of the black market.[56]

The number of international adoptions rose dramatically from 30 per year in 1989 to close to 5,000 in the period after passage of this law. Stories of baby bazaars began to emerge in the media.[57] They described families from the United States traveling to Romania, making contact with "baby brokers" and visiting institutions, where the physician in charge showed them infants available for adoption. Hard currency was used in the exchange, and liberal visa laws allowed families to return to the United States with their newly adopted children, often with very little understanding of what the children had experienced. At a moment in Romania's history when the people were quite poor and something had to be done about the thousands of children in institutions, international adoption to the United States and the United Kingdom was seen as extremely desirable.

The high rate of international adoptions led to negative media publicity and political attention and forced the Romanian government to amend its adoption policy. The government, in response, modified Law No. 11/1990. The new law, No. 48/1991, is often referred to as the "moratorium" on international adoptions. This outlawed private adoptions, established processes by which families were protected, and significantly restricted adoptions by families from outside the country.[58] The moratorium on adoption lasted for nine months.

The Romanian government vacillated during this period between restricting international adoption and expediting it. Authorities passed a law (47/1993) that facilitated international adoption. It did so by legally defining what it meant to be abandoned. The Judicial Declaration for Abandonment resulted in many more neglected children obtaining legal abandoned status and thus becoming eligible for adoption.[59] The Romanian Parliament then passed Law No. 84/1994, aimed at protecting the children, birth parents, and adoptive parents during the process of international adoption and preventing child-trafficking and other abuses. It also provided official recognition of an adoption from one signatory state to another.

Despite progress in the area of international adoptions, these measures did not reduce the rate of institutionalization of infants in Romania. This was due in part to the fact that Law No. 3/1970 was still in effect. This law required the government to place abandoned children in institutions, and in 1994, it remained the primary solution for children in struggling families. Even if a child were adopted, the odds were that he or she had previously been institutionalized.[60] Thus during this time period many children were institutionalized even with increasing international adoption. Poverty during this period (mid-1990s) was still rampant, and economic conditions remained dire.

NGOs from abroad worked with the Romanian government to help institutionalized children. They provided assistance, and often members of these groups spent time in the institutions.

The National Committee for Child Protection (NCCP) was set up in 1993. Even after its establishment, different institutions still reported to different government ministries, and abandoned children continued to be institutionalized.[61]

In the late 1990s, amid the efforts of international NGOs and the government to respond to the significant crisis in child protection, Cristian Tabacaru and Francois deCombret founded SERA Romania (Solidarite Enfants Roumains Abandonnes Romania), which would later become BEIP's major administrative partner. As SERA's general director, Tabacaru decided to introduce foster care in Romania as an alternative to institutionalization. Tabacaru hoped to implement a system of foster care similar to that used in France, where foster parents work for the state and receive a state-determined salary with a pension and other benefits. Nevertheless, his initial efforts to establish foster care were met with bureaucratic delays and skepticism that families would care for the kinds of children kept in institutions.

Tabacaru realized that he had to create the mechanisms through which foster care could become a real rather than a hypothetical alternative. As he explained:

> It's not enough to put in a law where a child could be protected—so-called—through foster care or institutionalization,

because if conditions are not in place for foster care, they will all go to institutions. It was as simple as that. (C. Tabacaru, interview, April 2011)

The November 1996 elections provided an opportunity for substantial reform. The Social Democratic Party of Romania, dominated by former Communist Party members, was defeated by the Christian Democrats, a center-right party.

The Romanian government passed a law (34/1998) that established the Department of Child Protection (DPC), which itself took the place of the National Committee for Child Protection. In March 1997, Tabacaru was appointed the first ever secretary of state for child protection in Romania.

Tabacaru's vision for child protection in Romania was twofold. First, he believed that children belonged in families rather than in institutions, and second, he felt that local control of child protection decisions was essential. These views were considered radical because of the huge bureaucracies that had grown to support centralized control of institutions for children. Many ministries in the national government received money for supporting the institution industry.

Nevertheless, Tabacaru convinced key government officials that it was less expensive to transfer management of child-protection cases to the local level than to continue handling them centrally. Locally, child-protection workers could combine and maximize resources, while remaining better informed about a family's needs. He also collected data indicating that it would be more cost effective to care for a child in a foster family than in an institution, on the basis of salaries at that time.[62] In fact, according to the World

Bank, in 1995 in Romania, the per-child expenditure for a young person living in a leagan was equal to the average wage of Romanian workers.[63]

The first step for Tabacaru was abolishing the former child-protection system. According to a newly passed law realizing Tabacaru's vision, foster care was preferred over institutionalization. Placing infants and young children into institutions was to be viewed as a last resort. Public services were restructured to provide help to abandoned children. These initiatives led to a decentralized system, with the local authorities and not the central government in charge of child protection.

Local authorities, who had detailed knowledge of individual cases, met about specific children and worked with social workers to recommend that a particular child either be placed in a foster home or returned to the biological family. The local authorities then decided the best way to proceed.[64]

Another aspect of Law No. 108/1998 was that infants and children could be adopted at younger ages than before. Biological parents could give up a child for adoption at birth but have thirty days in which to change their mind. The law outlined a process that had to be complete before adoption could occur. First, local authorities were to refer the child to the Romanian Committee on Adoptions. This group was then to inform all other county commissions to search for a Romanian adoptive family. If this search was unsuccessful, then the child became eligible for international adoption. International adoption with this process could take place in three months' time. This law (No. 108/1998) also transformed institutions. Leagane became "placement centers" for children of all ages with a range of physical and mental abilities.[65]

The law also attempted to establish a foster care system in which

Romanian families could take into their homes children who were abandoned or currently living in institutions. Gabriella Koman described the enormous challenge of recruiting and training a foster care workforce in a country that had no tradition of foster care:

> [They tried] to use foster families as an alternative to children put in institutions and to try to bring children from institutions within the foster network. It was more complicated for the age children more than twelve or thirteen years, both because of their, let's say, their long stay within the institution, but also the capacity of new foster families to take care of these children being affected by more than ten years in the institution. It was not easy. (G. Koman, interview, January 2011)

Resistance did not end with passage of the legislation, of course. In fact, one could argue it began then. Skepticism about Tabacaru, concerns about foster care, and resistance to change all played a part in the battle to promote foster care. The greatest challenge, however, came from the entrenched system being dismantled. In this case, the Ministries of Health, Education, and Labor all had substantial portions of their budgets committed to the institutionalization of children throughout Romania. Institutions raising children employed 70,000 people in Romania in 1999, and in rural areas, they were often the economic centerpiece of the village.[66] For example, Tabacaru found that the institution in Vistrate was ordering firewood at an alarming rate. Romanian winters can be harsh, but according to Tabacaru's calculations, if the institution in Vistrate burned a fire in every fireplace in the building, 24 hours a day, 7 days a week, 365 days a year, they would still use less wood than they had requested. When he dug deeper, he learned that many of the villagers were depending on the institution's supply for the wood they used in their homes.[67]

In October 1999, Tabacaru resigned his position in the government. By then he had already transformed the system, and he had created conflicts with some Romanian government officials, both for budgetary reasons and because his plan had decentralized child protection, taking authority away from the state and putting it into the hands of local authorities. In addition, the European Union (EU) opposed Tabacaru's pro-international adoption position. The first secretary of state for child protection then returned as director of SERA Romania, and in that capacity, he helped facilitate the launch of BEIP.

International Adoption: The Conflict and the Conclusion

One of the most publicized and contentious aspects of Romanian child protection in the late 1990s concerned international adoption. Aware of the widespread corruption in international adoptions, but also facing serious budgetary shortfalls that hampered efforts at decentralizing control over institutions, the Ministry for Child Protection decided to allow international adoptions because the NGOs and adoption agencies were willing to contribute financial support to the Romanian child-protection infrastructure.

Emergency Order 25 Regarding Adoption/1997 attempted to secure resources from adoption agencies to develop foster care and family support allowances. This strategy was seen as a way of addressing three major problems: 1) the criticism that international adoption prevented the development of domestic alternatives; 2) the limited resources available for decentralization and foster care; and 3) the corruption that had plagued international adoption since 1990. The law required agencies involved in international adoptions to pays fees to Romania for family and child services in exchange for access to children cleared for adoption. The contributions, which could be in the form of money, goods, or services, were assigned points at the local level and were reported to the central adoption authority (the Romanian Committee on Adoption). This committee then assigned the local agencies the right to place children legally freed for adoption on the basis of points they had earned through their contributions.

A second law authored by Tabacaru, No. 87/1998, addressed the conditions of adoption. The law outlined qualifications necessary for families to adopt a child in Romania and streamlined the process of domestic and international adoptions. The law mandated that "if a child is requested by both a Romanian and a foreign resident, priority is to the former." International adoption could not occur unless there had been an attempt to get a Romanian family to adopt the child. Despite the passage of laws to regulate international adoption, the "baby trade" continued on the black market through 2000. The Romanian Committee on Adoptions placed a moratorium on international adoption on December 14, 2000.

By 2001, it had become clear that the point system was not immune to corruption. Some adoption agencies were using it to obtain more children for international adoption, but funds were being misappropriated along the way. As a result, another moratorium on international adoption was enacted in 2001, just as the BEIP began.

This new moratorium was intended to allow Romania to develop additional safeguards for children involved in international adoption. New adoption and child-protection legislation was drafted with the input of international experts and submitted to the European Union for approval. However, Romania's efforts to reform its system of international adoption did not meet EU standards, and the moratorium was extended. In spite of the ban in 2003, more than 1,000 children were adopted by foreign families.[68] The government claimed that these were exceptional cases. The exceptional status granted to particular adoption cases allowed these adoptions to remain within legal boundaries while the moratorium was still in effect. This news generated fierce criticism from officials in the EU, supported by the threat to stop accession negotiations with Romania. The prospect of being denied admission to the EU ultimately led the Romanian government to outlaw international adoption in 2005.[69]

During the fifteen-year period from 1990 to 2005 the issue of international adoption and child protection was debated within Romania but under the shadow of a growing rift between the United States and the European Union. On one hand, the EU pressured Romania to severely restrict or even halt international adoption at several points during those fifteen years. The United States, on the other hand, had established itself as a supporter of international adoption, encouraging the practice and deploring extensive moratoria and the very restrictive legislation enacted in 2005 that essentially put an end to adopting Romanian children abroad. Although other Western governments, such as Italy, France, Spain, and Israel, where Romanian children had been previously adopted, sided with the United States, the official EU position was maintained.

Romanian government policy oscillated in response to competing pressures in order to 1) secure financial support from various international actors; 2) fulfill the country's aspiration to become a member of both NATO and the EU; and 3) avoid the major spending necessary to reform the child-protection system during the difficult transition from a welfare state to a market economy. As a result, Romanian government policies regarding institutionalization of children and international adoption were inconsistent and contradictory. While the government developed alternative forms of child care to facilitate deinstitutionalization of children, its policies on international adoption in the period between 1989 and 2004 varied in response to competing pressures from the United States and the European Union.

The EU position was that international adoption was corrupt, discouraged domestic adoption, and forestalled any effort to restructure the child-protection system. The EU parliament rapporteur on Romania, Baroness Emma Nicholson, an outspoken opponent of international adoption, insisted that the practice in Romania was a breach of children's rights:

> During the five years I spent as the European Parliament's Rapporteur on Romania it became clear to me that much of what was taking place in international adoption was merely a cover for a much deeper problem, which is a rampant and profit-driven free market economy in children. International adoption made it profitable for criminal gangs to coerce and deceive impoverished and unsupported mothers into giving up their children, who were then effectively sold on to western couples under the guise of "adoption." As this trade developed so too did the power of the criminal gangs at its centre.[70]

The United States, by contrast, acknowledged the problem of corruption associated with the practice of international adoption in

Romania, but proposed improving the system rather than ending international adoption. The U.S. government argued strongly that international adoptions could be supported without discouraging either reintegration or domestic adoption. It encouraged Romanians to focus on rooting out corruption.[71]

In the end, the EU position that equated Romanian international adoption with marketing children prevailed in dictating Romanian policy. There was little doubt about the ultimate outcome in this debate. Along with six other Eastern European countries, Romania joined NATO in March 2004, but it had not yet been admitted to the EU (accession occurred in 2007). This gave the EU enormous leverage, and, as EU rapporteur, Nicholson made it her mission to prevent international adoptions from Romania:

> I have no doubt the campaigners in the West who want Romania to resume *trading* in children are well intentioned. But I strongly believe the rights of children have priority and take precedence over any competing rights of adults. For example, the right of children to be happy and secure is more important than the right of childless couples to have children. I will make no apologies for campaigning to end this vile practice if it ensures that a child's future will be at the heart of his or her natural family, and I unreservedly celebrated when the bill banning ICA [International Child Adoption] was passed by the Romanian Parliament.[72]

Given widespread corruption, the EU position on banning international adoptions was understandable, but from the standpoint of children's best interests, the U.S. argument was compelling. In any case, the result of banning international adoptions was to eliminate one component of the child-protection solution in Romania.

Although this debate was unrelated to our project, which focused on Romanian families fostering Romanian children, because we were from the United States, some people wrongly assumed that BEIP was connected to the question of international adoption. Nicholson herself once made allegations linking the BEIP study to the corruption of international adoption. According to Gabriella Koman, Nicholson was convinced that the study was about assessing children for international adoption despite the fact that before and after an official inquiry, it was clear to Romanian officials that our study had nothing to do with international adoption. In the next chapter we turn to the foster care system that was at the very heart of BEIP.

Chapter 4

Ethical Considerations

Policy makers may need a strong method of proof to be willing to accept a given intervention, especially when it conflicts with their biases. Even if some experts claim to know, based on observational evidence or experience, that a particular intervention is effective in producing a given outcome, or is better than an alternative intervention, positive results from a well-designed randomized trial may be necessary to persuade policy makers to change standard practice.

Franklin Miller, "The Randomized Controlled Trial as a Demonstration Project" (2009)

It is hard to imagine a more vulnerable population than abandoned young children living in institutions. The risk of exploitation among these subjects—who cannot speak for themselves and lack advocates who are invested in them on a deeply emotional level—has been a major concern of BEIP from the very beginning. We have written extensively about the ethical dimensions of BEIP, and bioethicists have contributed several commentaries.[1]

Even the original discussions held within the MacArthur Network about developing the study design were heavily influenced by ethical considerations. As planning for the study continued, we discussed these concerns in greater detail as we learned more about the Romanian context.

Initial Considerations within the MacArthur Network

In research conducted with nonhuman primates and rodents, manipulating and controlling independent variables and experiences are relatively easy compared with the challenges facing those who conduct experimental research with humans, particularly children. With this in mind, we attempted to design the most scientifically rigorous study possible while maintaining the highest ethical standards. Much of the study design had to be shaped by which young children we could study—that is, their ages and circumstances, how long we could follow them, and the access that we could count on from the children's legal guardians.

We knew how we might examine the effects of early institutionalization on brain and behavioral development. There were, however, compelling reasons for setting up an intervention study within the context of a study of the effects of early experiences on brain development and behavior. First, if we confirmed our hypothesis that institutional rearing led to a derailment of typical development we would have lost valuable time in creating an intervention. Why not, then, we reasoned, implement the intervention at the outset? Thus instead of merely observing the effects of a potentially harmful environment, we could try to improve the chances for children's positive outcomes. This also makes the results of the study more relevant for clinicians, who are often faced with the dilemma of trying to help children who have experienced adversity to recover and adapt.

Second, given the concerns often raised about exploiting vulnerable children, we felt that it was far better for at least some participants to have a chance of benefiting from participation. Both researcher and subject stood to gain from this approach.

Third, an intervention study would be useful for those interested in developing child-protection policies both in Romania and abroad. Cristian Tabacaru was clear that a study of foster care was quite appealing because of the debates within Romania about institutions versus foster care as the best approach to abandoned children.

Fourth, embedded in an intervention study is a control condition that allows the researcher to observe the effects of *not* intervening—a major ethical issue in itself—which we discuss at length in this chapter. By comparing children raised in an alternative approach to "care as usual," it becomes possible to gauge the improvement that the intervention provides.

Finally, from a scientific perspective, intervention studies have the potential to shed light on underlying mechanisms—in our case, a foster care intervention could permit us to draw inferences about *why* institutionalization is bad for children. For all these reasons, an intervention study seemed scientifically and ethically preferable to a study that merely documented the effects of institutionalization on abandoned children.

Selecting the Intervention

The design that we selected for our intervention, a randomized controlled trial (RCT), is not without controversy in the context of institutionalized children. From a scientific perspective, it is the most desirable way to study the effects of an intervention, and the only way that allowed us to argue that the intervention is causally related to outcomes. But the notion of randomly assigning children to continued institutional care or to placement with a foster family gave us pause. In the United States, foster care had been the preferred method of caring for orphaned, abandoned, and maltreated children for nearly 100 years, and certainly since the 1960s or so, foster care had been virtually the only approach. The literature certainly seemed to suggest that early life in an institution was debilitating.

We had all seen the films Rene Spitz made of institutionalized children in the 1940s (http://www.youtube.com/watch?v =VvdOe10vrs4), and we all had a strong bias that foster care was a better approach. At the same time, we were unaware of any study that had ever used a randomized controlled trial design to compare the two. In fact, when we looked at the literature in 1999, we found only seven studies published in English comparing young children in foster care with young children in institutions. In all these studies the children placed in foster care had developed more favorably than the children in institutions.[2] Though many of the studies had significant methodological limitations, at least by late-twentieth-century standards, the consistency of their findings was notable. Yet not one of them had used random assignment. Thus there was simply no way of ruling out the possibility that in previous studies, the more handicapped children had remained in institutions and the better functioning children in families. This could plausibly explain why in those studies, children placed with families seemed to fare better than the children placed in institutions. From the standpoint of our ethical considerations, the fact that there had never been an RCT on this issue increased our justification for undertaking one now.

Ethical Deliberations within the BEIP Core Group

The Romanian Context

We were aware, as we discussed in developing the design of the study, that within Romania, there was no agreement about whether institutional care should be perpetuated or foster care implemented in its place (see Chapter 3). Romania came under increasing pressure during the late 1990s and the early 2000s from the European Union, and to a lesser extent the United States, to develop alternatives to caring for children in large institutions. Nevertheless, within the Romanian child-protection community itself, there was much suspicion about foster care and more support for the status quo.

In April 2001—just as we initiated the intervention— Charles Zeanah went to dinner with Sebastian Koga, BEIP project manager, and with the directors of the six institutions for young children in Bucharest. The dinner had been arranged to encourage and thank the directors for their cooperation in the study that was soon to begin. The director of St. Andrew's Placement Center, Dr. Elena Tarţa, was very proud of the way her institution ran and the care the children there received. She had become the director just before the fall of Ceauşescu in 1989 and had designed what she considered to be a model institution in Sector 1 in Bucharest. Throughout the dinner, she argued that foster care was dangerous for young children and that institutional care, when administered properly, was preferable to it. This was not the first time we had heard this argument from someone who worked in child protection. In fact, a former U.S. ambassador to Romania, James Rosapepe, noted in his final report to the American government that in

Romania, many child-protection managers and employees often failed to understand or support the national strategy of developing alternatives to institutional rearing.[3] Like Dr. Tarţa, they believed rumors about harsh treatment of children, pedophilia, and organ-trafficking in foster care. Some of the justification for institutions was that the complex developmental problems from which many children suffered required expertise that professionals but not "untrained" foster parents could provide.

Furthermore, there was a belief among some that young children in institutions were already damaged. As described in Chapter 3, in the Ceauşescu era, children had been assessed and selected dichotomously as normal or "irrecuperable" at age three. Those deemed normal would continue in typical institutions while those considered irrecuperable would be sent to medical institutions with terrible conditions. In this line of reasoning, young children were defective *before* being placed, and the developmental delays and deviant behaviors that they displayed led to rather than resulted from institutional rearing. In fact, a significant minority of the children had serious genetic or metabolic syndromes that left them handicapped, probably reinforcing this view. In our opinion, however, the notion that young children were destined to serious developmental problems and that intervention was pointless reflected an outdated model positing that children's development unfolds relatively impervious to experience. We believed that all of the children could have benefited from a more responsive environment, and many of them would have had minimal to no problems if they had been raised in more normal circumstances.

A genuine uncertainty about the best form of care for orphaned, abandoned, and maltreated children prevailed within Romania, and especially in Bucharest, which in terms of child-protection reforms lagged behind some other parts of Romania ten years after the revolution. Having decided—at least preliminarily—on an RCT, we then began to consider in detail other ethical issues involving the project.

Do No Harm

One of the first principles we considered was how to ensure that our intervention and the study designed to evaluate it "did no harm," at least to the extent that we could control potential for harm. We quickly chose foster care as the intervention to study as a logical alternative to institutional rearing, instead of trying to improve institutional care in some way. Although foster care is not entirely without risks, we had good reason to believe that living in a foster family would be less harmful for children than living in an institution, since all previous studies had demonstrated that children in foster care had better outcomes than children in institutions; moreover, from both policy and scientific perspectives, foster care represented a dramatic contrast to institutional care.[4]

Principle of Noninterference

Another way we attempted to reduce any potential harm was by ensuring that we did not interfere in any way with children's placements *because of* their participation in the study. We planned to remove children from institutions and place them in foster homes, of course, but we did not want to interfere otherwise in their placements. That is, if children were adopted or returned to their parents, we would not interfere in any way. Romanian law was very clear on these points; we adopted these ethical principles even before we learned about the law.

We made one exception to this rule: we sought to ensure that no child who had been removed from an institution would ever have to return. Because the financial support for our study was limited, we were concerned about initiating an intervention, collecting data, and concluding the project, with the end result that children who had been placed in project foster homes might then be returned to institutional care. This seemed to us to be an important part of the "do-no-harm" doctrine, though we were uncertain about how much control we would have over these issues. We tried to negotiate this arrangement with each of the local commissions on child protection within Bucharest. Of the six sectors with institutions for young children, all but one agreed to this condition.

Unfair Advantage

One could argue, of course, that it is unfair to provide a putatively enhanced experience (that is, placement in a family) for half the children, while others continued to live in institutions that were on their face objectionable to us. Nevertheless, we reasoned that no child would be maintained in an institution *because of* participation in the study. In fact, if the study had not been conducted, all of the children would have continued to live in institutions.

Minimal Risk

Another tenet of the "do-no-harm" principle, particularly with a vulnerable population, is to use measures and procedures that pose no more than minimal risk. Using vulnerable populations to test risky interventions or invasive measures is unacceptable, of course. The U.S. government defines minimal risk as when "the probability and magnitude of harm or discomfort anticipated in the research are not greater in and of themselves than those ordinarily encountered in daily life or during the performance of routine physical or psychological examinations or tests."[5] We also wanted to be sure that we were using measures with which we were familiar and comfortable.

Of course, Dr. Tarţa and others in Romania considered foster care itself to be a risk. Nevertheless, foster care was a legal method of child protection for abandoned children in Romania and had been for several years before our project began. Although media reports regularly document abuse of children in foster care in the United States, the actual incidence is low, less than one half of 1 percent of children in foster care are abused or neglected, a figure that is lower than the roughly 1 percent of children in the population at large who are abused or neglected.[6] We found no available evidence to suggest that foster care *increased* risk over institutional care. Therefore, we concluded that the risk of foster care was not an impediment to conducting the research.

Stop Rule

In studies of new drugs, if preliminary results indicate that a particular drug is clearly more effective than a placebo, the trial can be stopped and the drug made available to everyone, even though the RCT has not been completed. Unfortunately, we realized even before we started that a stop rule would not be an option for us because of the high cost of foster care. We simply did not have the funds to remove the sixty-eight children randomized to care as usual from the institutions and place them into foster care. Roughly 50 percent of the budget for BEIP, all supported by the MacArthur Foundation Research Network on "Early Experience and Brain Development," went toward foster care—salaries for the fifty-six foster parents, expenses for the children in their care, and salaries for the three social workers who oversaw the foster care network. As we will describe later, when our initial data began to demonstrate advantages for children placed in foster care, we decided on another approach to address the inequity of the two arms of the intervention.

Investigator Bias

Beyond study design, we also needed to protect the study from our own biases, to the degree that it was possible. Measures of brain functioning, EEGs, event-related potentials (ERPs), and MRIs, as well as neuropsychological measures (for example, CANTAB, Cambridge Neuropsychological Test and Automated Battery—see Chapter 6) and biological measures (for example, genetics), provided no opportunity for bias, as these data do not involve caregiver report or observer bias. For growth measurements and psychological tests such as the Bayley examination or the WISC-IV (Weschler Intelligence Scale for Children), the possibility of bias is always present, although both the administration and the interpretation of results are fairly standardized. More vulnerable to bias are ratings of observed behavior during laboratory procedures and caregiver reports. If observers knew, for example, that they were watching an institutionalized child or a child who was living with her family and had never been institutionalized, that knowledge might affect how they perceived and rated the child's behavior. For assessments of attachment, emotional expression, and interactional behavior, in particular, bias could pose potentially serious problems. To minimize the influence of bias, observations were videotaped and later coded by raters who were unaware of group membership of the children and, in some cases, unaware of the study design. This effort was aided by a rule that all institutional caregivers wore street clothes during all procedures. The trained raters who coded these videotapes were told only that they would be coding young children in homes and in group care settings. They were as "blind" as any coders of any study using behavioral assessments could be.[7]

For many of our measures, we obtained reports from the caregivers in the institutions, and during the follow-up we often compared these reports with those from the foster caregivers or biological parents (if children were reintegrated). There is no foolproof protection against caregiver or parent bias about children's behavior and characteristics. On the other hand, when possible we compared both observation and caregiver reports on particular aspects of behavior (such as indiscriminate behavior) and often found a high concordance between these two data sources. This increased our confidence that the caregiver reports were valid. Furthermore, we also considered the pattern of findings across all measures and determined that those completely removed from the possibility of bias, such as EEGs, yielded results similar to those that had more possibility of bias, such as attachment behavior in the Strange Situation Procedure, described in Chapter 6.[8] We are satisfied that the results obtained so far from the BEIP have been similar across all types of measures, and we are confident that bias did not greatly affect the results that we reported.

Ethical Evaluation of the BEIP

Ethicists typically evaluate the soundness of research projects using specific indicators. We have grouped the issues into five categories: social value, risk/benefit ratio, informed consent, independent review, and post-project contributions. Despite considerable overlap among many of these constructs, they each yield distinctive considerations.

Social Value

There is widespread agreement that the first ethical test of research is whether it has societal value.[9] If a study has no value then it is unethical to have people participate, even if the risks are minimal. At first, this seems straightforward, of course, but it quickly becomes more complicated. That is, a study cannot become ethical only *after* it is completed; it must be ethical prospectively.[10] So how were we to judge the social value of BEIP before we had even conducted the intervention? And who must find it valuable?

It's fair to ask whether we already knew the answer to the key question we were posing. After all, we shared a bias about the advantages of foster care, and the literature comparing the two interventions indicated that in ten out of ten studies, children in foster care were developing more favorably than their counterparts in institutions. Although an RCT had never been conducted in this context, RCTs may not be necessary or even desirable to evaluate interventions.[11] For example, in areas of food insecurity it would be unethical to randomize which children are fed adequately.[12]

Clinical equipoise, which some researchers consider a minimal ethical standard, refers to a genuine uncertainty among professionals, before conducting a trial, about whether one intervention is superior to another. In the United States, one would be hard-pressed to find child-protection professionals or child-development researchers who had any doubts about whether institutional care or foster care is the superior intervention; a few academics, however, have actually advocated vigorously for a return to orphanages because of their putative advantages over foster care.[13] Even in the

United States, then, there is not unanimity regarding optimal placement settings. In Romania, the question about institutional rearing was even more unsettled.

So was BEIP socially valuable? In our view, there were several reasons that it was. First, the weight of the evidence on the question of institutional rearing and foster care rearing was limited and clearly susceptible to biased results. From an ethical perspective, the study would have been harder to justify if a large number of randomized controlled trials worldwide supported foster care over institutional care. The policy and public health implications of the results have enormous implications for the millions of children around the world being cared for in institutions, and this begs for a solid evidence base. Second, we were not proposing to conduct a clinical trial that deprived young children of the opportunity to be raised in foster care. We were instead introducing a new and promising intervention in a setting in which the standard of care was institutional rearing. Third, the secretary of state for child protection in Romania, Cristian Tabacaru, thought the study was not only acceptable but also necessary in order to convince others in the Romanian government that there were alternatives to institutional care.

But what about clinical equipoise? Franklin Miller and Howard Brody argued convincingly—in our view—that "clinical equipoise is neither necessary nor sufficient" for conducting ethical RCTs. They assert that clinical equipoise confuses a standard of physician-patient relationship with the conduct of a clinical trial. A physician treating a patient is obligated to do what is best for that individual patient. An investigator is obligated to the larger society that stands to benefit from the results of this research.[14] By randomizing children in BEIP, we necessarily allowed half of the children to be in an intervention—institutional care—that we considered suboptimal, but we could not prove that it was suboptimal without conducting the study. And without the randomization, there would be no benefit to anyone other than the children who actually participated.

Risk-Benefit Ratio

Related to social value is the balance between risks and benefits of study participation. Every study has both risks and benefits for participants, and investigators are obligated to ensure that the benefits or potential benefits outweigh the risks or potential risks. How did we evaluate the potential risks and benefits of BEIP?

One may ask why we were conducting the study in Romania rather than in the United States. We had been invited there, of course, in part because of a policy debate within Romania, but we were also very sensitive to the fact that we were U.S. investigators studying an exceptionally vulnerable population in a fledgling democracy. It is important that investigators from more developed countries not conduct research in less developed countries merely because they are able to exploit lax regulations or access more unprotected populations.

First, we went to Romania because there were tens of thousands of abandoned children being raised in institutions there. The study could not have been conducted at home because so few young children in the United States are raised in institutions. Although a few infants and toddlers are housed in "congregate care" settings in the United States, the placements are usually short-term and the numbers of children involved quite small. According to data from the Child Welfare League of America, less than 1 percent of the children younger than three years of age in out-of-home care in the United States are in institutions, and this includes young children requiring intensive medical support.[15] According to data from 2004, across all 50 states, the District of Columbia, and Puerto Rico, fewer than 1,500 children under three years of age were housed in institutions (meaning group care staffed with less than full-time parents), and many had severe disabilities. Further, whereas Romania had 205 institutions with more than 100 children each in 2001, the United States had none.[16]

Second, and more important, BEIP was aimed at enhancing the health of abandoned children raised in institutions, and the results of the study were likely to benefit other, equally vulnerable children. Benefiting populations similar to those being studied is in keeping with guidelines published by the Council for International Organizations of Medical Sciences (CIOMS).[17] In addition, government officials and child-protection advocates, like Cristian Tabacaru, had asked questions about policy that a methodologically sound scientific study could address. By demonstrating whether removing young children from institutions and placing them in foster care led to developmental gains, we could show whether a change in policy *in Romania* was warranted or not. Further, the government could potentially consider the economic, psychological, medical, and social risks and benefits of different approaches. In particular, we had an opportunity to provide evidence supporting or refuting the widespread belief that "orphanage children" were permanently damaged or "irrecuperable."

We also considered the fact that we were not asking any child to experience prolonged institutional care *because of* participation in the study. As a result of Romanian law and our noninterference rule, children who were adopted or returned to their parents were simply followed in the study, but their actual placements were unaffected by study participation. In fact, by the time the intervention arm of the study was completed when the children reached fifty-four months of age, only twenty children continued to live in institutions, whereas forty-five lived in project-sponsored foster homes, twenty were living in government-sponsored foster homes that did not exist at the time the study began, eighteen had been returned to their biological parents, two had been returned to biological relatives, and sixteen had been adopted.[18]

When early returns from BEIP indicated clear advantages for those children in foster care, we were forced to reconsider the "stop rule." Because we lacked the resources to make foster care available for all children, we decided to communicate our results to those who might make use of the data. We organized a press conference to announce results of the investigation and invited the relevant ministries in the Romanian government to attend. The United States ambassador to Romania at the time, Michael Guest, introduced our project and each of the three investigators at the press conference on June 13, 2002. Some government officials attended, but we did not know how they reacted. Gabriela Coman, secretary of state for child protection from 2000 to 2004, told us later that she was very aware of our study and findings at the time.[19]

In 2002 and 2004, we also organized and funded (with support from the MacArthur Foundation, the National Science Foundation, and the Harris Foundation) two national conferences in Bucharest on child development at which we presented our preliminary findings. We invited health and mental health practitioners, child-protection professionals from the government and NGOs, neuroscientists, and government officials to attend and asked many to speak at these meetings. In addition to contributing to the Romanian child development and child-protection infrastructure, these conferences were another way to make research results from BEIP available as quickly as possible to those who might make the best use of them.

The conferences we organized are relevant to a larger question of benefits—namely, benefits to society. If risks to participants are low, as they were in BEIP (without the study all the children would have experienced extended institutional care), then advantages that accrue from knowledge gained from research become an important consideration. Although much progress has been made in developing alternatives for abandoned children in Romania in the decade after Ceauşescu's overthrow, thousands of children remained in institutions there (according to data from the European Union, in 2012, approximately 8,000 children, many of them handicapped, were still living in institutions). The results of BEIP have many potential benefits for the larger population of institutionalized children in Romania, and perhaps those in other parts of the world, as well.[20] There are millions of abandoned children worldwide, many of whom are being raised in institutions.[21] Policy-makers concerned with how best to enhance the long-term outcomes of abandoned children should attend closely to research comparing different forms of care. Results from BEIP, with appropriate cross-cultural and contextual cautions, have implications for institutionalized children worldwide.

Informed Consent

With regard to informed consent, the critical questions focused on who was legally responsible for children who were wards of the state, and who would advocate on their behalf. Measures for protecting those who cannot consent include ensuring that they are enrolled only in minimal-risk research and that the results of that research benefit either the participants directly or the population from which they were drawn. Judged by the ethic of "fair subject selection," BEIP met the standard in our view.[22]

But the question remained, who could consent for institutionalized children? Longitudinal child-development research projects and RCTs, such as BEIP, were unprecedented in Bucharest in 2000, and we could not find any established procedures for ethical review of the study within Romania. At the same time, Romanian law outlined procedures that we were required to follow. By law, children cared for in institutions were in the custody of the state. The legal guardian of a child who was a ward of the state was usually the Commission for Child Protection for the sector/county in which the child resided (sometimes the mayor). Commission members were professionals from the community who were appointed by the mayor; they generally included a physician, the City Hall secretary, a representative from the Ministry of Labor, the director of child protection, and a representative from the police. The commissions, of course, were legally established entities with no connection to the investigators or the investigation. Thus, in the first step, the commissions had to agree to allow specific children to participate. Following Romanian law, the commissions reviewed the placement of each child in state custody every three months. On the basis of their recommendation, children in the institutionalized or the foster care groups could be returned to their families or adopted within Romania. Participation in the BEIP did not limit or affect in any way removal of children in the institutional group from institutions and their return to biological families, domestic adoption, or placement in the government foster care program that developed as the project continued.

From the outset, the commissions required us to obtain a separate consent from the birth parents of institutionalized children whenever possible. Interestingly, they required this consent not so the child could participate in the study but so he or she could be placed in foster care. As noted, most institutionalized children in Romania had at least one living parent. In these cases, though children were in the legal custody of the state, parents retained some rights. Some parents demanded institutionalization for their children, and they were allowed specifically to preclude the government from placing the children in foster families. According to Bogdan Simion, executive director of SERA Romania, this decision was sometimes based on the biological parents' fears that the child, if placed in a family, would begin to "prefer" another mother figure. For us, it underscored the degree of suspicion with which foster care was regarded at the beginning of the study.

As an additional protection for children, we decided to ask institutional caregivers (or foster parents) to "assent," or agree to, the specific activities and procedures involved in the study. We reasoned that because these caregivers/foster parents knew the children best, they might be more psychologically invested in them than were the distant government officials who were their actual legal guardians. So caregivers for institutionalized children and foster parents could always decline or terminate any activity or procedure that they deemed too unpleasant or undesirable for the child. For example, several young children objected to wearing a cap containing electrodes for recording EEGs and ERPs. Research assistants made determinations about whether to persist, but caregivers accompanying the child always had the final word about terminating the procedure if a child was distressed.

The situation was more straightforward for children living with their biological parents, who gave permission for their children to participate in the study and assented to specific procedures. The initial consent forms that we developed in conjunction with three different University Institutional Review Boards (IRBs), however, were quickly judged to be overly long, legalistic, and ultimately inappropriate in the Romanian context. To ensure that consent was truly informed, the commissioners requested shorter, more explicit consent forms, which we developed and had approved by the IRBs at our academic institutions.

Independent Review

Review of BEIP occurred on several different levels. Tulane University (Zeanah); the University of Minnesota, Boston Children's Hospital, and Harvard University (Nelson); and the University of Maryland (Fox) each had IRBs that reviewed and approved the entire protocol. Our Romanian collaborators at SERA were unaware of a comparable system for review of research projects in Bucharest at the time the project began, and as a result, the study was approved by the local Commissions on Child Protection in Bucharest, as well as by the IOMC, a clinical entity operated by the Ministry of Health in the Romanian government. It was important to our Romanian partners at SERA and the IOMC that the study be conducted within the principles outlined by the United Nations Convention for the Rights of the Child.[23]

In response to questions about BEIP raised by a European Union parliamentarian, in June 2002 the Romanian government appointed an ad hoc Ethics Committee, comprising academics and government officials familiar with child development and child-protection policies. This group visited the BEIP laboratory and reviewed all aspects of the project in detail, concluding that it was ethically sound. More recently, we worked with Bucharest University administrators to develop a formal institutional review board modeled after those at U.S. universities.

Post-Trial Obligations

Many believe that researchers have obligations after a project is completed. According to CIOMS (1993), researchers ought to ensure that interventions resulting from the research are made "reasonably available" to the community. An alternative to this approach, called the "Fair Benefits" framework (2002), suggests that benefits to the community may comprise many different types of advantages, including ancillary care, training of health care personnel, employment, and economic stimulation, as well as the intervention being made available after the trial. We have attempted to fulfill our post-trial obligations, with varying degrees of success, in several ways.

The central idea in our original planning was that no child randomized to foster care would have to return to institutional care. This was in part a concern for the child and in part an attempt to make a lasting contribution to the child-protection system in Bucharest. From the beginning of our intervention in April 2001 through the formal completion of the intervention in 2005 (when the youngest children reached fifty-four months of age), no children were returned to institutions except for one, who received an emergency placement for two weeks when the single foster mother caring for the child died suddenly. Two weeks later, when the BEIP social worker learned that this had occurred, she immediately identified an alternative home and had the child placed there. By the conclusion of the project, the local governmental sectors within Bucharest had assumed administration and support of all the MacArthur foster homes in which children were placed, substantially increasing the numbers of government-supported foster homes in Romania's largest city.

Unfortunately, since the trial ended there have been a few lapses from our original agreements. Some children with severe developmental delays have been placed in institutions for the handicapped, and when some foster parents have become seriously ill, the children have been returned to some form of group care. In many of the institutions for older children, students attend public school "off-grounds." Some children also live in "social apartments," which are located within typical apartment complexes. They are staffed by part-time caregivers and house four to six children. As more serious behavior problems have developed as the children enter adolescence, some have been placed in institutions analogous to group homes or residential treatment centers in the United States.

We and our partners at SERA have advocated vigorously on behalf of these children, but the limits of our influence have become painfully apparent. Adina Codres, former director of St. Catherine's, at one time the largest institution for young children in Romania, said that the global financial downturn of the mid- to late 2000s hit Romania hard and resulted in serious budget cuts within child protection. In fact, she noted that the National Authority for Child Protection was incorporated into the Ministry of Labor in 2010, reflecting both a smaller budget and weakened authority.[24] Bogdan Simion of SERA Romania agreed that the budget cuts have led to reductions in the availability of foster care in Bucharest.[25] In 2010, for example, salaries for foster parents were cut 25 percent.

Beyond the study itself, we have made other attempts to contribute to the Romanian infrastructure. For example, our Romanian staff, led by Sebastian Koga, was awarded a USAID grant to conduct training in child development throughout Romania. From

2004 to 2005, they conducted trainings for child-protection workers and other professionals throughout Romania on topics such as fostering attachment, encouraging language development, and managing behavior problems.

A final but important legacy of the project has been the creation of a Bucharest-based Institute for Child Development (IDC), which opened formally on October 31, 2011. BEIP was awarded a grant from the John D. and Catherine T. MacArthur Foundation to partner with SERA Romania and, as appropriate, various government agencies to create the IDC, which is intended to continue training, research, and service delivery to at-risk children and serve as a resource for the entire country.

Our hope is that the IDC will establish the infrastructure and critical mass of local expertise necessary to respond to the changing needs of the nation's vulnerable child population in a self-sufficient, effective, and sustainable manner. The idea is to enhance the knowledge infrastructure on child development in Romania to equip the country's child advocates, policy-makers, and clinicians with up-to-date models and approaches to deal with future needs. Another goal of the IDC is to provide the Romanian child health and welfare community with the sound scientific information necessary to make good policy decisions. This includes widespread dissemination of relevant educational and training materials and emerging research findings to community practitioners, parents, key government officials, the media, and the general public. A final goal of the IDC is to provide evidence-based, scientifically driven care to post-institutionalized and special needs children, to improve the identification of children at risk, and to generate strategies for effective early intervention that can be implemented throughout the country.

Conclusion

From the beginning, the BEIP was conceived as a scientific and humanitarian effort, with the goal of balancing those considerations. We recognized from the outset that the ethical dimensions of the project were complex and deserved considerable planning. We have been aware, too, that some people may find certain aspects of BEIP objectionable on ethical grounds. Most though not all reviews by bioethicists of BEIP have been favorable.[26] Because we cannot imagine that the children in the study could have been better off had we not conducted the study, we are satisfied that it was ethically sound and worth conducting. Still, we wish that the policy implications of BEIP were more fully appreciated within Romania and beyond, especially with regard to the quality of foster care needed for young children who have experienced adversity. We consider the larger implications of BEIP more fully in Chapter 12.

Chapter 5

Foster Care Intervention

> There's no way government can raise kids. But government
> can do things that help support parents who are raising kids,
> and government can also be the safety net for the poor and
> vulnerable children who, for whatever combination of rea-
> sons, are not being adequately cared for by their own parents.
> Hillary Rodham Clinton (March 3, 1996)

BEIP was designed to address a number of essential questions: Can a carefully planned foster care program enhance the development of children who have been abandoned early in life and placed in institutions? And would the timing of this intervention, the age at which children were taken out of the institution, or the duration of the intervention itself, have an effect on outcome? Having decided on foster care as the intervention to evaluate in the Bucharest Early Intervention Project, we were confronted with how best to ensure high quality in a context in which foster care was uncommon, re-garded with suspicion by some, and largely unused as a method by the local child-protection system. To do this, we drew upon our experiences implementing an intervention program for maltreated children in New Orleans.[1] We wanted to create a foster care pro-gram in Bucharest that was appropriate in Romania, affordable, replicable, and informed by developmental science. If successful, we hoped that we might provide a model alternative to institutional-ization for abandoned children.[2]

As described in Chapter 2, BEIP was implemented at a time of great flux in the organization of child protection in Romania. Reform legislation had passed in the mid-1990s, but many parts of the country had not yet felt its effects, creating an opportunity for us to provide data relevant to the debate among those in the reform movement who wanted radical changes and those who preferred more modest changes to the system. We begin here with an overview of foster care for abandoned children in Romania in the post-Ceauşescu era, concentrating on Bucharest and contrasting practices there with those in other parts of the country.

Although Romania has received the most media attention for its child-protection problems, its challenges are shared by most of Eastern Europe and Russia, as well as by countries in many other parts of the world.[3] Millions of dollars in resources from the European Union and the United States, in particular, have been provided to aid in reforms, and progress is undeniable, though the problems are far from solved and, in many cases, stubbornly impervious to change. Controversies abound in any discussion of contemporary Romanian child protection, and though Romania is moving inexorably toward more family-centered approaches to abandoned children, the end is not yet in sight.

Foster Care in Bucharest

In the Ceauşescu era, foster care was legal, but it was rarely used in Romania as an intervention for abandoned children. Informal fostering among relatives occurred, but government-sponsored foster care was rare through the 1990s in Bucharest. In fact, this was true throughout Russia and Eastern Europe, for several reasons.[4] First, in countries where a large portion of the population faced financial and housing difficulties of their own, many families found it hard to consider caring for someone else's child. Lack of government assistance only exacerbated this problem. Second, cultural prejudices about children who had lived in institutions, and prejudices about Roma children in particular, which persist in Romania, further discouraged foster care.[5] Third, limited public awareness of foster care as an entity kept potentially interested families unavailable. Finally, the absence of a tradition of using nonrelative foster care or even adopting children who were not biological relatives may have contributed to rumors that such systems promoted harsh treatment of children, pedophilia, and organ-trafficking.

Although government-sponsored foster care in Bucharest was quite limited in 2000, NGOs involved in international adoption had established foster care there in the 1990s. Some of these agencies promoted reunification or domestic adoption, such as programs in Constanţa and Bucharest supported by Holt International, though many were more interested in supporting international adoption.[6] Typically, they identified abandoned children who were eligible for adoption and living in institutions, and then placed them in foster care. After several weeks to months, the children were adopted directly from foster homes. Two factors contributed to the existence of NGO-sponsored foster care.[7] First, in the point system created by Tabacaru (see Chapter 3), NGOs that contributed to the child-protection infrastructure, even by providing foster care, received credit toward access to children eligible for international adoption. Second, some families from abroad interested in adoption preferred children who were in foster care to those in institutions. For this reason, the length of time a child spent in foster care was often short.

The foster care system in Romania at the time was modeled after the French system, in which foster parents, or "maternal assistants," were hired as full-time employees of the county (or sector, in Bucharest) and paid a salary with benefits. This situation differs from that of foster parents in the United States, who receive only a modest subsidy for each child in their care.

According to Bogdan Simion of SERA Romania, other areas in Romania had made earlier and more substantial commitments to foster care than had Bucharest.[8] For example, by October 1998, the remote northern county of Bistriţa in Transylvania had made foster care a viable option to institutionalization.[9] Just a year after passage of the new law authorizing foster care, six foster families were caring for children, and three years later the county had sixty-eight licensed and active foster families, with another fifty licensed but without children because of funding limitations. According to Marin Mic, a social worker who directed child protection for the county of Bistriţa at the time, the number of children living in four institutions in the county fell from 850 to 220, and most of those who remained in institutions were severely handicapped children or older adolescents. The number of children in foster care increased to 220.[10] The work in Bistriţa was the result of a partnership between the Bistriţa County Department of Child Protection and an NGO, the Romanian Children's Relief/Fundatia Inocenti, for whom Mic now works.

In contrast to Bistriţa, Bucharest saw little immediate change in foster care availability after 1997. Attention seemed to be more focused on disentangling sector responsibilities. Exact numbers are elusive because records were less than systematic at that time, but three years after passage of the 1997 legislation to increase foster care, only a handful of foster homes were operating in Bucharest. Effectively, foster care in the capital city was not a part of a child-protection continuum of available services at the time. Perhaps because the institutions in Bucharest were large and in some cases old (St. Catherine's, founded in 1897, at one time housed 1,000 children), they may have been even more entrenched and wary of change than facilities in more remote areas like Bistriţa. According to Adina Codres, director of St. Catherine's from October 1998 to June 2000, physicians and caregivers at St. Catherine's were well aware of the debates among government officials about the merits of foster care versus institutional care, and many feared for their jobs as a result.[11]

Social Work in Bucharest and Romania

We recognized that given the lack of foster care in Bucharest, a vital component of creating a foster care network would be recruiting and training highly skilled social workers to support and monitor the children and families in the study. The unusual history of social work in Romania made these goals especially challenging.

In March 2001, when Veronica (Vera) Proaspatu was twenty-five years old, she received a phone call from a former classmate, Alina Rosu, a social worker in Bucharest. Alina had been hired as the first BEIP social worker in January, and she wondered if Vera might also consider applying for a job as a social worker. Vera had had nearly two years of experience with institutionalized children during her days as a student, when she worked full-time in Sector 5 in Bucharest. As a social worker there, she was responsible for assessing and counseling children and families, and representing children's interests at meetings of the commission. She was also responsible for designing case plans for children, attempting, whenever possible, to reintegrate them into their biological families.

Although still a student at the time, Vera was hired for this complex work because Sector 5 had suddenly changed the minimum requirement for social workers from a fourth grade (primary school) education to a high school (eighth grade) education. They did so because the law that Tabacaru had written in 1997 *required* the local commissions on child protection to assume responsibility for all the children born in their sector. In the past, Sector 5 officials had relied primarily on institutions in Sector 1 to care for these children, but now they had to manage on their own. Young social workers like Vera were hired and strongly encouraged to return children to their families whenever possible because the cost of caring for these children was now assigned to Sector 5.

Vera's previous experiences with institutionalized children, combined with her work with street children the year after she graduated, made her one of the most experienced social workers in Bucharest we could have hired when we launched BEIP. Although social work has a long and proud history in Romania, Ceauşescu had closed schools of social work throughout the country in 1969, on the premise that, under Communism, no one was supposed to have social problems. Only twenty years later, after the revolution, did the school of social work in Bucharest, now a part of the University of Bucharest, reopen.[12] The first undergraduate degrees in social work after the revolution were granted in 1994, and within five years the university had added a masters degree program that took a year and a half to complete. In 1999, the social work school became an independent entity, the Faculty of Sociology and Social Work, distinct from the Faculty of Psychology and Educational Sciences. By 2000, Romania had seven universities with departments of social work that were graduating 500 social workers a year.[13]

Although the number of social workers was increasing, Ceauşescu's decision to close schools of social work in 1969 meant that an entire generation of senior social workers was lost. New graduates, for their part, were left without mentors or traditional practices to follow in the decade after the revolution. Similar to psychology, Romanian social work also had been out of touch for decades with progress in the field at large.[14] Vera's professors during her undergraduate and graduate years in social work school included sociologists, physicians, economists, police officers, and even psychologists, but no social workers, other than one field instructor.

At the time Alina Rosu called Vera to try to interest her in BEIP, we were convinced that the social work staff that would implement and nurture the foster care intervention was an essential key to its success. Therefore, despite the unique problems facing the social work profession in Romania, we wanted to hire employees who were enthusiastic, eager to learn, and able to innovate. Fortunately for us, Vera accepted the offer, and a third social worker, Amelia Greceanu, was hired in June 2001 to complete the team. All three of the new social workers embodied the characteristics we sought, and they enthusiastically approached their new and challenging tasks.

Guiding Principles for the Foster Care Intervention

As we designed the training and established a structure for working with our social work team, we knew only that foster care was rare in Bucharest and that institutionalization was clearly the preferred form of care for abandoned children. So in developing BEIP, we relied on our experience working with young children in foster care in the United States.

Anna Smyke and Charles Zeanah had worked since 1994 in a community-based intervention in New Orleans that attempted to provide comprehensive mental health services to young, maltreated children.[15] This work included intensive efforts with children and their biological parents and children and their foster parents. Anna had worked extensively with foster parents, attempting to build successful relationships with children in their care, who would remain with them for about eighteen months before either returning home or being freed for adoption. As a result, she was well aware of the challenges to a healthy relationship that could arise from the child, from the foster parent, or from the unique fit between the two. Although she had not had experience with institutionalized children in New Orleans, she was quite familiar with the effects of serious social neglect. Therefore, with involvement and input from the three of us, she constructed and oversaw the implementation of the foster care intervention. We used guiding principles, derived from experiences in child protection in New Orleans and our reading of the literature, to inform the training of the BEIP social workers as well as the foster care program we planned to create.

Caregiving Quality

We believed that the key to a successful foster care intervention was for parents to provide sensitive care to the young children assigned to them. This meant getting to know the child as an individual and emotionally investing in him or her, in the context of a long-term commitment.

One of the striking differences in the foster care we sought to establish and the foster care we knew in the United States had to do with duration: in Romania foster care would not necessarily be a temporary placement. For young children in the United States in foster care, states require the implementation of a "permanent plan" for their custody within about eighteen months of placement. That is, they are supposed to be either reunited with their biological parents, ideally because the parents have remediated the problems that led to their child's entering the system, or freed for adoption, following formal legal termination of parental rights. Foster care is intended to be a relatively short-term intervention. In Romania, in contrast, there was no push to resolve children's custody, and legal pressure to terminate parental rights was rare in our experience, especially since international adoptions had been banned.

In essence, we were advocating that foster parents care about and for the children as if they were their own. Evidence demonstrates that high-quality foster care parenting is essential to adaptive behavior, well-regulated emotions, and the physiology of young children who have experienced maltreatment.[16] This approach, focused on providing enhanced *experiences* for the child, made the parent-child relationship the central component of the intervention, in keeping with contemporary research and practice.[17] For the social workers, this meant learning to think about parent and child characteristics that might impede, limit, or diminish a close and loving parent-child relationship, as well as ways to intervene to enhance the relationship.

Adequate Support

We were also aware that one of the most frequently cited frustrations for American foster parents is lack of support and assistance from the social workers assigned to them. Welfare systems in the

United States are chronically underfunded, and one consequence is that case loads are often too high and the paperwork burden too great. In addition, in our experience there is often little effective training provided to child-protective services workers regarding how to establish, maintain, and manage relationships between case workers and foster parents. If the worker and parent like each other, all goes well, and if not, there is much stress and conflict. We were determined not to leave these vital relationships to chance but to help the social workers understand how to be responsive and how to bring a therapeutic rather than a managerial approach to their relationship with foster parents.

We also believed that it was crucial to provide adequate support for the BEIP social workers, who were engaged in a pioneering expedition that had little precedent in Bucharest. We planned to have supervisor/consultants in the United States who would work with them on a regular and sustained basis to maintain fidelity in the approach, adapt to unforeseen challenges, and help them feel supported in their challenging assignments.

Comprehensiveness

We learned early that developmental and especially behavioral resources were scarce in Bucharest at the time we began the project. The Institute of Maternal and Child Health, directed by Alin Stanescu, a prominent pediatrician in Bucharest, ran a community clinic to which we referred children for speech and language therapy, motor problems, and global developmental delays. But we could not identify mental health professionals with early childhood expertise to whom we could refer children. This meant that our team had to be prepared to provide interventions for typical behavioral and regulatory issues that present in young children—oppositional and aggressive behavior, social withdrawal, toileting difficulties, bedtime struggles and nightwaking, and refusal to eat or hoarding, hyperactivity, and indiscriminate behavior, all of which are common in young, maltreated children.[18]

Cultural Sensitivity and Importability (Transportability)

We were mindful of the risks of implementing a model of parenting for Romanian foster parents that was imported from the United States and not culturally appropriate in Bucharest. We were fortunate to learn about a foster care training manual developed by World Vision that had been written by and for Romanians modeled on similar manuals in the United States. Our administrative partners, SERA Romania, had had several years of experience implementing foster care in other parts of Romania. We vetted our training and consultation model with them, as well as with our own social work staff, to ensure the cultural appropriateness of the approaches we were implementing.

Another important issue involved the economic disparity of salaries in Romania and in the United States. We wanted to ensure that we paid salaries to foster care parents that were commensurate with the importance of their job. At the same time, these salaries needed to be reasonable in the Romanian context for several reasons. First, from an ethical perspective, we did not want to create undue coercion for parents to participate in the study. Second, from a practical standpoint, we did not want people to "do it for the money," as is sometimes alleged about foster parents in the United States. Finally, at the conclusion of the project we wanted to have created something affordable that could be replicated. We consulted both with SERA Romania and with sector officials to ensure that salaries and other reimbursements were fair and consistent with established standards in Bucharest.

Recruiting and Training Foster Parents

Recruitment

Vera recalls recruitment as the first major challenge of her new job with BEIP. Foster care was so unknown in Bucharest that it was difficult to identify prospective foster parents. We followed recruitment strategies for foster parents similar to those used in the United States, including newspaper advertisements and posted flyers, and Alina Rosu did several radio interviews. Local commissions had to certify parents before they were eligible to be paid, and this included verifying educational attainment, employment, and absence of criminal histories. At the time our project began, completing gymnasium (eighth grade) was required, but one of the foster mothers had been certified when only completion of elementary school was the minimum requirement. BEIP social workers organized groups for prospective foster parents to help navigate them through the process of obtaining a certificate from the sectors. For most, this was an unfamiliar, and at times intimidating, process.

These early efforts helped the social workers to gain a better understanding of the foster parents and to anticipate appropriate matches between foster child and foster parent. We sought foster parents who seemed comfortable with young children and who seemed to possess the emotional availability necessary for the intensive work of caring for a post-institutionalized infant or toddler. These were qualities that the team of social workers assessed as they interviewed and observed foster parents during the training.

Because foster parents are full-time employees in Romania, one person is designated the legal "maternal assistant." In each case in our program the officially designated foster parent was a woman. All of the women were ethnically Romanian, and their average age was forty-seven years, with a range of twenty-nine to sixty-five years. The majority of the foster mothers (twenty-nine of the fifty-three) were married and living with their spouses; eleven were widowed and nine divorced; and four had never married. Some of the mothers had fostered children previously (twelve of the fifty-three). In fact, eight had fostered ten or more children in the past, and two of those eight had fostered twenty or more. Their experience was not in government foster care, of course, but rather in foster care provided by international adoption agencies. Given that such programs had only existed for four years, many of the mothers who had had many children had had them for very short times.

Ten years after the project began, Vera did not recall any differences in the commitment or skills of foster mothers who had had previous experience compared with those who had not. A majority of the foster mothers had their own biological children (forty-two of the fifty-three), but most (twenty-three of the fifty-three) had only one biological child. Of the mothers with no biological children, only one had had experience as a foster mother, meaning nine women had had no previous experience as mothers. All of the mothers stated that they practiced their religion, and all but one were Christian Orthodox.

Training

The training provided to foster parents was conducted over six weeks. It included twenty-two hours of instruction on legal issues, such as children's rights, the recent reforms in Romanian child protection, and the rights and obligations of foster parents. Another twenty hours covered child development, including information from educational and child psychology, but no mention of attachment, which was a cornerstone of our approach. Finally, eighteen hours was spent describing challenging behaviors and conditions in children such as disabilities, misbehavior, social challenges, and some suggestions about how to cope with these.

Our work in the United States has taught us that though such trainings may be generally useful, foster parents often find themselves ill prepared and confused about how to understand or respond to a child's behavior. What they learned in class is not adequate preparation for actually parenting a child. In such situations a social worker is crucial to helping the parent understand the child's behavior and respond effectively rather than impulsively.[19]

Additional BEIP Training for Foster Parents

Foster parents visited with their prospective foster children while they were still living in institutions. This enabled foster parents to gain an appreciation of the child's institutional experience and begin to establish a relationship with the child. Three visits were scheduled. Most of these were held in the institution, though our social workers always attempted a visit in the community, for example, in a park or a playground. Most institutions refused to allow these outside meetings, however.

Visits were also a way of assessing the match between the foster parent and the child. For example, as anticipated, four foster parents declared that they did not want a Roma child placed with them. These declarations were honored, though one of the four who had expressed reservations agreed to foster a child when the ethnicity was in doubt. When it was later confirmed that the child was Roma, she agreed to keep the child.

From the outset, the social workers emphasized the importance of attachment, meaning that parents should fully invest in the child's well being, work to understand the child's needs, and respond appropriately. Urging parents to commit fully—meaning accepting the child as their own for the foreseeable future—was a radical approach in Bucharest at the time, as it might be even now in many places.

Placing the Children

The bureaucracy of the sector commissions proved an unanticipated barrier to placement of the children. Naïvely, we had assumed that with our study blessed by the National Authority for Child Protection and the Ministry of Health, with money to pay foster parents, and with foster parents legally certified, the process would proceed smoothly and quickly. Not so.

Meetings of the Sector 1 Commission on Child Protection took place in the director's office, in the administrative building of St. Catherine's. These meetings were remarkable scenes that we witnessed more than once. All of the people in Sector 1 who wanted or needed to appear before the commission showed up on the day of the meeting. People might want to see about obtaining benefits for a period of time (for example, a monthly allowance or daycare), to get a handicap certificate, to place their child in an institution, or to make a request for adoption. We could always tell when a meeting was scheduled because 50–100 people were milling around outside the administrative building of the placement center. No one had a specific appointment time—everyone was on a list for the day and waited until their names were called; at that point, they were permitted to appear before the commissioners. If their name was not called, they were reassigned to another day's list and hoped to be called then.

Once called, the petitioners entered a big office in which the commissioners sat around three sides of a large table, facing those who sought their help. Seated behind a stack of papers, many of the commissioners smoked continually, and several conversations occurred simultaneously. The dim lighting and the smoke-filled air created a kind of luminous haze in the room that added to the surreal quality of the experience. The petitioners sat in front of this august body and made their requests.

The BEIP social workers became familiar attendees at these meetings, patiently waiting their turn, often for hours, and sometimes without ever being called, to appear before the commission. When they did appear, they dutifully presented the required paperwork, but they often found that the month's requirements included information or documentation that had not been mentioned in the previous meeting. Sometimes the commissioners themselves wanted additional documents and sometimes the sector social workers, who often attended these gatherings, suggested that more paperwork was necessary. This created additional delays. Interestingly for us, the commissions all required that the biological parents of the children in BEIP, assuming their whereabouts were known, consent to having their child placed in foster care. In the case of the study, however, they did not require parental consent. Instead, they considered that, as the child's legal guardian, the commission itself could consent to study participation without consulting the biological parents. We understood this to mean that they were sufficiently skittish about foster care to want an extra layer of protection. Vera recalls the challenge of convincing some of the commissioners that we were not "selling" the Romanian children abroad.

This process prolonged placement delays. On average, it took a little more than two months between baseline assessment and actual placement in a foster home, with a range of same day to six months. The first four children were placed on April 11, 2001. One child, originally randomized to foster care, never was placed with a family because her birth certificate was determined to have been fabricated. The police advised the commission to watch the child closely, and they decided that she must remain in the institution.

For the most part, one child was placed with one foster mother, but four sets of siblings were kept together, and six other foster mothers agreed to care for two children. Five children randomized to foster care were never placed in foster care. Three of these were reintegrated with their biological families immediately after baseline assessment, and two more were adopted by Romanian parents immediately after baseline assessment. Thus, fifty-three foster parents were hired initially to care for the sixty-eight children randomized to the foster care group (see Table 5.1).

Providing Support

Financial Issues

As noted, Romania had adopted the French system of foster care in which one parent was made a salaried employee. Following SERA Romania's experience and advice, we initially paid each designated foster parent an average of 200 RON. (RON was the Romanian currency that preceded the Euro; in 2001 200 RON was roughly

Table 5.1 Children placed in foster care

Foster mothers (53)	Children in FCG (68)
4 foster mothers cared for 4 sets of siblings	8 children
6 foster mothers agreed to care for 2 children each	12 children
43 foster mothers cared for one child each	43 children
	5 children randomized to foster care never placed in foster care (2 adopted and 3 reintegrated with biological parents immediately after baseline assessment)

equivalent to $77 USD per month.) We also provided a monthly subsidy of 50 RON (roughly $19 USD). We gave some credit for amount of education and previous experience as a foster parent. To put this in perspective, the gross minimum wage in Bucharest in 2001 was 133.3 RON per month, and the average gross wage was 422 RON per month. For parents who cared for two foster children, the monthly salary was 1.5 times the amount of those who cared for one child, and the subsidy was doubled.

When the social workers placed a child in a foster home initially, they typically supplied the foster family with a bed for the child, some clothes, and a baby carriage. The social workers also provided medication or other items as the need arose.

Pediatric Support

We contracted with a pediatrician in Bucharest to provide care as needed to the children in the study. Although the physician was also available to children in the care-as-usual group, because they were mostly in institutions on units headed by physicians, they rarely if ever saw the pediatrician. As a group, however, the children in foster care made between one and four calls or visits a week to the pediatrician, at least in the initial phase of the intervention, though the calls declined over the course of the project.

Caring for Young, Post-Institutionalized Children: The Intervention

We recognized that the challenges for foster parents asked to care for post-institutionalized infants and toddlers were likely to be considerable. Thus the BEIP foster care program would require extensive support, both for the foster parents working directly with the children and for the social workers helping the foster parents. Each of the BEIP social workers was responsible for roughly one-third of the study participants in foster care and their foster parents. Following placement of the child, foster parents in the BEIP network received frequent visits from the social workers. Initially, families were visited weekly. After several months, if the child was settling in, visits were every other week for several more months and then monthly after about a year. This frequent contact was designed to ease the transition into families for children who had been raised in institutions and to help the social workers detect and intervene early if problems arose. Foster parents called the BEIP social workers as needed throughout the project.

Challenging Behaviors among the Children

Although most children navigated transitions from institutional care to family care successfully, some required special support. These children displayed one or more of the problems listed in Table 5.2. Early in the placement, a common challenge for foster parents, for example, was the children's "loudness." Many foster parents lived in large apartment buildings that were part of Ceaușescu's legacy. These were constructed without much soundproofing, and children's loud and disruptive behavior was a particular concern for foster parents. The social workers helped the parents understand that the change from an institutional environment to a home was significant, and that children needed some time to adapt. Bath time was a particularly unpopular event in the institutions, as it usually involved impersonal scrubbing and hosing down of the children with little regard for their comfort. Not surprisingly, foster parents had to gradually introduce bathing as a fun activity to desensitize the children to what was obviously a troubling memory. When fos-

Table 5.2 Behavior problems reported by foster parents following initial placement

Regulatory	Developmental	Behavioral	Emotional	Social	Other
sleep	cognitive delays	hyperactivity	reticence	indiscriminate behavior	motor stereotypies
eating	language delays	loudness	anxiety	social withdrawal	
toileting	motor delays	agitation	crying spells	aggression	
		oppositional behavior			

ter parents struggled, social workers spent more time with them and increased the frequency of home visits.

Challenges among the Foster Parents

Individual differences in children were one thing, but the social workers also began to notice many differences among foster parents. Some called frequently to discuss their children, and others rarely called. Some seemed to have appropriate developmental expectations for young children, others did not. Those who had fostered previously often felt sadness about children who had been in their care and were later adopted out of the country. Some foster mothers, particularly if they were older and experienced, were defensive and not open to learning from our team. For example, one retired teacher insisted that she knew all about children, from being a mother herself and from teaching hundreds of children in her career. She had two young foster children placed with her and insisted on initiating toilet training immediately. The social worker urged her to give the children, both of whom were less than three years old, some time to settle in. The foster mother was offended that her approach was questioned. She was also unhappy that the children did not respond to her efforts. Despite repeated attempts on the part of the social workers to engage with her, she never accepted a collaborative problem-solving approach. One of the children in her care was returned to her biological parents, and the other, by mutual agreement, was moved to another BEIP foster home.

To address issues of parenting, the social workers provided many hours of one-on-one counseling with the foster mothers. In addition, with help from their supervisors, the social workers created a voluntary support group for foster parents. Several foster parents were experiencing unresolved grief about children they had previously fostered, who often were adopted suddenly, with little time to prepare. Despite concerns about privacy and speaking up, the foster mothers who attended seemed to benefit from talking in a group about their experiences and provided support to one another. The group seemed to help foster parents appreciate that others had similar concerns and experiences and provided them with concrete suggestions for managing their young foster child's behavior.

Challenges from the System

Despite reforms in child protection, the bureaucracy continued to thrive. A large source of stress for our team was negotiating with sector social workers. They were, in Vera's words, "paper focused" rather than "child focused." The notion of the child's "best interest," which is important if inconsistently applied in American child welfare departments, was a foreign concept to sector social workers. They displayed no particular urgency, were quite devoted to formal processes, and seemed firmly committed to rules and regulations. Over time, the BEIP social workers noted, their professional relationships with the sector social workers improved, especially after the sector employees attended two large conferences that we organized in Bucharest. Nevertheless, despite improvements in individual relationships, the BEIP social workers saw little evidence that the system was changing.

Work with Biological Parents

Although BEIP was envisioned as a foster care intervention, not all of the children remained in foster care. Some were adopted, and some were returned to their biological families. The social workers also provided counseling, support, and parenting interventions to biological parents whose children were returned to them. Most biological parents were known to the BEIP social workers because they had worked hard to locate them to request consent for foster placement for the child. Remarkably, in some cases, parents said that they had not even known where their children were. Many were pleased to receive information about the children. If the sector commissions decided that they could be reunited, we assisted in the reunification plan as needed. By the time the children were eight years old, thirty-one of them had been reintegrated with their biological families.

In addition to the typical challenges associated with transitions, Vera recalls that many of the foster parents did not want to talk with the children about what was happening to them. The social workers encouraged families to explain to these preschool children, using simple language, where they had been before coming to live with them and why. This kind of discussion was meant to help children understand their situation, be less fearful about more sudden placement changes, and to know their own histories.

Consultation, Supervision, and Support for Social Workers

Supervision as a professional process was virtually unknown in Romania when we first began our collaboration. Some have drawn the distinction between "administrative supervision," which is about supervisors checking that activities have been completed according to protocol, documented properly, and so on, and "reflective supervision," which is focused on supporting the clinician in understanding her own and her client's behavior.[20] The supervisor, who is ideally less directly involved than the clinician and more experienced, provides support and guidance, but also encourages understanding behavior and reflecting on one's own responses. The social workers had experienced only "administrative" supervision in previous settings, and this was the model they brought to BEIP.

The model that we implemented was one of reflective consultation/supervision. We expected the work with foster parents to be challenging and complex, and we were committed to helping our Bucharest team to develop the equanimity to handle the stresses involved.

The supervisor/consultant was Anna Smyke, an applied developmental psychologist who had years of experience working with young children in foster care and their foster parents at Tulane University. Anna was involved from the outset in developing the intervention, and she oversaw its implementation throughout the first four years of the project.

Anna began her consultation by meeting with the team in Bucharest; she then had weekly video or phone consultations with the staff. Initially, these were training sessions about early childhood development, attachment, and the challenges of foster parenting, but gradually the discussions became more focused on questions about cases. Most of her contact with the social workers was through the weekly video or phone consultations.

In June 2003, roughly two years into the project, Valerie Wajda-Johnston, a clinical psychologist at Tulane, joined the team as a supervisor. She also began by meeting the team in person followed by weekly video or phone consultations. This second layer of supervision was added to provide additional support by focusing primarily on developing and implementing behavior programs with families and handling challenging behavior among foster parents. These sessions followed a similar trajectory to Anna's; that is, they began with more formal presentations on topics of interest to the team, but increasingly they became case-based discussions. These discussions included what might be hindering a particular intervention, whether a specific intervention was appropriate, and how to alter the intervention or change interventions.

Barriers to Consultation/Supervision

Effective communication between the team and the supervisors was initially challenging. Although the social workers had to be competent in English to be hired, they did not necessarily feel comfortable speaking in English, particularly about unfamiliar topics, including abstract psychological constructs. The sessions began as telephone calls, but this made establishing a working relationship more challenging. Once video-conferencing was implemented, and as the sessions became routine, the social workers seemed more comfortable.

Psychological barriers also had to be identified and overcome. Although the social workers were eager and curious, they also seemed reluctant at first to "present" their cases—even informally —to Anna or Valerie during supervision sessions. They were especially reticent when it came to talking about their personal reactions to the work. What eventually emerged were issues of trust. The social workers worried about the purpose of these consultation/supervision sessions. They wondered if supervision was actually some kind of disguised evaluation of their job performance rather than an opportunity to get help to better understand clinical dilemmas. They wondered if what they said could somehow jeopardize their positions.

Another challenge was that many of the ideas and practices the supervisors discussed with the social workers were unfamiliar to them. The team sometimes felt uncomfortable about implementing ideas they had not seen in practice. But having limited experience themselves and no models for this kind of work in their training or work experiences, the social workers had little choice but to try various approaches, relying on their own intuition and the experiences of their supervisors.

Once they identified these challenges, the group found them easier to discuss and overcome. For example, the social workers realized that the foster parents' reluctance to be open with them and seek help was similar to their own initial reticence during consultation sessions with their supervisors.

Child-Protection Outcomes

This book devotes considerable attention to the outcomes of BEIP (Chapters 7–11). These are child-centered outcomes, meaning they determine the degree to which specific aspects of brain functioning and a wide range of behaviors can be altered by altering children's experiences. Here, we consider another type of outcome more relevant to those involved with systems of child-protective services.

Did foster care actually provide better-quality caregiving than institutional care? Quality of caregiving was significantly better for the children in foster care than for the children in institutions at both thirty and forty-two months of age, and in fact, it was indistinguishable from quality of caregiving in the families of never-institutionalized children from the community.[21] In addition, we wanted the foster care intervention to be safe and stable for the young, post-institutionalized children, who had already experienced serious deprivation. During the four years of the intervention phase of the project, we detected no instances of maltreatment or mistreatment. In addition, 84 percent of the children remained in their original BEIP foster placement until the intervention phase ended (generally when the child reached fifty-four months of age). Nine disruptions did occur: three children were removed from two different foster mothers who became ill; one foster mother died suddenly; two foster mothers requested removal of the children because they could not cope with their special needs; one child was transferred to be with a biological sibling; one foster mother decided to take another job; and one child was removed because the match between the child and parent was not a good fit. In each case, we continued to follow the children in the study.

As described in Chapter 3, we arranged for the foster care program that had been supported by the project to be transferred by the local governmental sectors in Bucharest, as we had negotiated at the outset of the project. Because Romania was undergoing a transition from institutional care to foster care at that time, the local governments hired BEIP foster parents and used governmental social workers to monitor the needs of these families. The BEIP-sponsored network of foster care was transferred to the Romanian local authorities at the conclusion of the intervention, when the child was fifty-four months of age. Thus the transfer process occurred over about three years because of the two-and-a-half-year age difference in the oldest and youngest child in the study. By the time the children had turned eight, thirty-five of the original fifty-six foster parents continued to care for BEIP children.

Current Status of Foster Care in Romania

Much has changed since we began our work in Romania in 1999 and 2000. Cristian Tabacaru points out that 30,000 foster parents throughout Romania now belong to a union—an unimaginable prospect for most in the mid-1990s, when he conceived and wrote the reform legislation.[22] Institutions are no longer the preferred option for caring for abandoned infants. In fact, in 2004 the Romanian government passed Law 272/2004 (section 2, Placement, article 60), which made it illegal to institutionalize a child before his or her second birthday, unless the child was severely handicapped. The preference for foster care in Bucharest and throughout Romania is an amazing systemic transformation in a mere fifteen years that should be appreciated and, in our view, celebrated.

At the same time, the status of foster care in Romania as we write today remains deeply troubled. The worldwide economic downturn of 2008 and after led to serious budget cuts in Romania. Some foster parents have had their salaries cut by 25 percent, and an article in January 2011 noted that 1,000 foster parents throughout Romania had resigned because of these salary cuts.[23]

Gabriela Coman, former secretary of state for child protection in Romania and now director of the National Center for Missing and Sexually Exploited Children in that country, fears that the child-protection situation will deteriorate considerably if the economic situation does not soon improve. She pointed out that in January 2011, not only were foster parents resigning in large numbers, but also a governmental hiring freeze was in place that made it impossible to recruit and train new foster parents.

A critical issue from our point of view is whether the salary approach to foster care is sustainable in a country that is struggling economically as much as Romania is. The alternative is a subsidy system like that used in the United States and many other countries. Of course, most people would agree that child protection is underfunded in the United States as well, so perhaps whatever system is in place will experience funding difficulties because abandoned and maltreated children will never constitute a powerful constituency.

The lesson from BEIP is not merely that foster care is better than institutional care, though we do believe that is true. Rather, it is that a high-quality, child-centered foster care program that focuses on helping foster parents commit emotionally to the children in their care is better than impersonal institutional care. We are troubled by reports from BEIP social workers that the child-protection system in Bucharest is still, in their opinion, often driven more by bureaucratic processes than by children's needs. This situation is hardly unique to Romania, of course, but until it changes, much remains to be done.

Chapter 6

Developmental Hazards of Institutionalization

> It is recognized among workers in education and in child
> psychology that children who have spent their entire lives in
> institutions present a different type of their own and differ in
> various respects from children who develop under the con-
> ditions of family life.
>
> Anna Freud and Dorothy Burlingame (1944)

In 2010, as many as eight million children around the world lived
in some form of institutional care.[1] The reasons for institutionalization vary by country and context: some children are abandoned at
birth owing to social and religious pressures (for example, un-
wanted pregnancy and religious restrictions against abortion), others suffer from poverty or neglect (including physical and sexual
abuse), and many more are orphaned as a result of war or disease.
Although institutions for children exist in most countries, not surprisingly they are more prevalent in areas of extreme poverty, or
those having experienced war or health epidemics such as HIV.

Institutions for young, typically developing children are virtually
nonexistent in the United States and Great Britain. In the early part
of the twentieth century psychologists in these countries identified

two issues that changed social attitudes toward institutionalization as an option. The first involved psychoanalytic approaches to understanding personality development that gained influence in the first part of the century. With them came a view of the importance of maternal-infant interaction as a foundation for subsequent social development. By the middle of the century John Bowlby, a British child psychiatrist and psychoanalyst, argued compellingly that maternal deprivation early in life had significant consequences for the emergence of psychopathology. In his paper "Forty Four Juvenile Thieves," Bowlby presented case studies of children who were living in institutions as a result of delinquency. He noted early adversity in the lives of these children and emphasized maternal deprivation (either early abandonment or maternal neglect). He concluded that experiences of early maternal separation compromised subsequent personality development.[2]

At the same time in the United States, another child psychoanalyst, William Goldfarb, published a series of papers on children being raised in institutions, describing atypical social and personality development that he attributed to maternal deprivation.[3] During World War II, Anna Freud observed the social behavior of young orphaned children rescued from the Holocaust and living in Great Britain. Her observations of their atypical social behavior confirmed the growing sentiment among psychoanalysts that separation from the mother at an early age was detrimental to social personality development even in the presence of adequate physical care and peer interaction.[4]

Anna Freud's observations were followed soon thereafter by the studies of the American psychoanalyst Rene Spitz, who described the effects of leaving infants in hospitals for long periods of time. He provided dramatic film and behavioral evidence for the negative effects of abandonment and routine hospital care on young infants. Spitz coined the term "hospitalism" to describe the effects of lack of stimulation and absence of social interaction on infants who were kept for long periods of time in hospitals and other institutions without contact with their mothers.[5]

In the wake of WWII, which left so many children orphaned or abandoned, the World Health Organization commissioned a report by John Bowlby on the mental health consequences of institutionalization of children. The report Bowlby wrote emphasized the detrimental effects of institutionalization, and specifically maternal deprivation, on children's social development and mental health. It influenced social and mental health policy, particularly in the United States and Great Britain.[6]

A second issue psychologists identified in the mid-twentieth century was the significant effects that institutionalization had on cognitive and motor development. Psychologists studying children were interested in determining which features in the environment shaped learning in the early years of life and, conversely, what the outcomes were for young children denied these basic patterns of stimulation. Reports suggested that infants raised in institutions had delayed motor and cognitive development. One study in the United States found that removing infants from institutions and placing them in a home for women with intellectual disabilities resulted in dramatic improvements in motor and cognitive skills, no doubt a result of the enhanced attention ("enriched experience") the infants received from their new caregivers.[7] Ironically, infants who were in institutions because they were orphans were used as "practice babies" by departments of home economics in the United States. Universities such as Cornell would bring these babies to the campus, where students would learn to care for them.[8]

This practice, the basis for the novel *The Irresistible Henry House,* was discontinued in the early 1960s.

Yet a third area emerged as important in the second part of the twentieth century, this time from the domain of neuroscience. Studying the impact of experience on brain function and development, neuroscientists distinguished between experience-expectant and experience-dependent changes, as described in Chapter 1.[9]

Experience-expectant changes are those that occur as a result of the "expectation" of certain experiences common to all members of the species (for example, patterned light, complex sound, consistent caregiving). Psychologists in the 1960s enumerated the types of stimulation and experiences that they found important for typical infant and child learning, which included vestibular, kinesthetic, and tactile stimulation and contingent social interaction. (Contingent social interaction can be thought of as "serve and return," that is, the well-coordinated turn-taking that occurs when parent and infant interact with each other.)[10] Infants and young children in institutions develop in environments in which these experience-expectant stimuli are absent, hence potentially creating a weak foundation to brain development in the first years of life.

Experience-dependent changes are also important in the context of institutional rearing. For example, some researchers have speculated that indiscriminate social behavior may be adaptive in settings of social deprivation where children may seek social contact wherever they can find it. Nonetheless, there is clear evidence that those children who exhibit high levels of indiscriminate behavior while living in institutions are at increased risk for psychiatric impairment two to four years later.[11]

The results of studies showing significant effects of institutionalization on young children's physical, social, intellectual, and mental health therefore have raised important scientific and policy issues. What factors, types of stimulation, or circumstances were necessary for typical development in young children? How could these factors be implemented within the context of care for abandoned and neglected infants and young children? How early in life was it necessary to provide these factors or circumstances and for what period? Could change be implemented within the context of institutional settings, or was it necessary for the physical and mental health of the children to remove them entirely from this environment? These issues set the context for the research undertaken in the BEIP, and many of them continue to be debated today. In this chapter, we explore research on the effects of institutionalization.

The Context of Child Care in Institutions

Historical descriptions of orphanages and the development of children reared in them have painted a bleak picture of institutionalization.[12] Indeed, given such accounts, which were particularly numerous after WWII, it is no wonder that child-protection movements in both the United States and Great Britain eliminated institutions for young infants under almost all circumstances. In 1962 Sally Provence and Rose Lipton, describing an institution in the United States, wrote of how children remained inside nursery rooms for much of the time, often kept in their cribs for most of the day and night, and experienced human contact primarily for instrumental purposes such as diapering and feeding. Children of the same age were housed together and were tended to by overworked caregivers who changed and bathed them with little time for play or social interaction. Feeding consisted of propped bottles with formula to which pureed food was added as children grew older. Although sometimes as many as twenty children of similar ages were placed together, they had little interaction with one another, and they rarely spent time outside or away from the institution.[13]

Understaffing and frequent turnover meant few opportunities for infants to be held or to experience meaningful interactions with their caregivers.[14] One child might be cared for by up to thirty individuals who worked rotating morning, evening, and night shifts.[15] Provence and Lipton observed that rooms were eerily quiet given the presence of so many infants. Apparently, because no one responded to them, young infants learned not to cry when distressed. Other children appeared to be unresponsive to auditory stimuli, again because it carried little meaning.[16] Feeding, changing of diapers, and being picked up by a caregiver just happened to these babies; nothing they did elicited the interaction.[17]

So what exactly were the consequences for the cognitive, language, and socioemotional development of these children? An infant expects an environment in which he or she will receive stimulation and interact with responsive caregivers. But what happens when these basic expectations are not met?

Head Circumference, Height, and Weight

We begin with an examination of the effects of institutionalization on physical growth. The relations between early institutional deprivation and stunted growth have been examined in detail in the English Romanian Adoptees study (ERA).[18] Initial effects were dramatic, with many of the Romanian adoptees measuring over three standard deviations below the developmental norms on head circumference (a measure closely associated with brain size) at their time of entry to the UK. These effects were strongest in those who experienced the longest periods of deprivation. From early childhood to adolescence there was considerable catch-up, especially in those most affected. By age eleven, head size was nearly normal in the group with less than six months of deprivation; but in those experiencing longer periods of deprivation, head circumference was still substantially below developmental norms. Researchers also studied the effects of subnutrition, as indexed by weight at the time of entry into institutions. Subnutrition was defined as being 1.5 or more SDs below the UK growth norms. Many institutionalized children (though not all) fell below this figure.[19]

Cognitive Development

Over the years, many papers have documented the effects of institutionalization on the cognitive development of children. These reports can be divided up into those that examined children living in institutions, those that compared children in institutions with those adopted out or returned to their biological families within the same country, and those that have followed institutionalized children adopted internationally. In one meta-analytic review reporting on seventy-five studies across three types, the authors found that children reared in institutions had consistently lower IQ scores compared with their typically developing age mates.[20] The authors found that children living in institutions assessed when they were younger than four performed less well compared with children assessed after their fourth birthday. Age at entry into an institution played an important role in determining IQ level of the children. Children placed into institutions before twelve months of age did less well than children placed at a later age. The authors did not find an overall significant effect of length of stay in institutions. Nor were differences in caregiver-child ratio between the institutions significantly related to differences in effect sizes for IQ. Even the contrast between the most favorable caregiver-child ratio (maximum of three children per caregiver) versus ten to one in other orphanages did not lead to a significant difference.[21]

Thus while institutional care had a definite effect on IQ, the reasons for this effect, other than age of entry into the institution, were not evident from the meta-analysis. In part, these results are a function of the great variability in caregiving contexts across the seventy-five studies, as well as differences in the reasons children are placed in institutions. But it does point to gaps in our understanding of the effects of institutionalization on IQ in young children. In the following section, we review the three types of studies outlined above and highlight gaps in our knowledge that are answered in the BEIP study.

IQ of Children Living in Institutions

More than fifty years ago, Wayne Dennis and Pergrouhi Najarian described significantly lower IQ scores for forty-nine Lebanese children living in a foundling home (or orphanage) as compared with a community sample. This pattern of findings has been reported numerous times by other researchers.[22]

More recently, just after the collapse of the Ceauşescu regime, a study was conducted on young children living in institutions in Romania. Researchers studied twenty-five children living in a leagan in Timisoara, a region northwest of Bucharest. Results of the intellectual assessment showed that the orphanage children were severely delayed, with none functioning at age level. Twenty of the twenty-five were functioning at levels less than half their chronological age. Using a number of standardized developmental measures, the authors reported no correlation between Apgar score (an estimate of overall function obtained minutes after birth) and a child's Bayley score (the Bayley Scales of Infant Development is a standardized assessment of development that has age norms); nor was there a correlation between length of time in the institution and Bayley scores.[23] These findings echo the results of the meta-analysis described above with regard to no relation between length of time in an institution and IQ.

A recent study from 2010 examined children living in institutions in Ukraine.[24] The authors compared the cognitive development of HIV-infected children living in institutions with noninfected children also living in institutions, as well as with HIV-infected children living with their families and typically developing children living with their families. Their data suggest that institutionalization, more than HIV infection, was responsible for severely low IQ scores. Both the HIV-infected and the noninfected children living in institutions had IQ scores between 60 and 70 (severe retardation), while HIV-infected children living with their families had a mean IQ of 79, performing better than noninfected institutionalized children, who had a mean IQ of 69.[25]

A number of studies with institutionalized children have involved interventions attempted while the children were still living in these institutions. One such study completed in Romania in 2005 involved two intervention trials, one in 1991–1992 and the second in 1993–1994 in a leagan in Iasi, Romania. Findings from both studies showed significant intervention effects in social behavior and IQ. The intervention appeared to delay decreases in IQ, but it did not provide a significant enough boost to enhance IQ scores to those of typically developing community controls.[26]

In 2008 a study by the St. Petersburg–United States of America Orphanage Research Team examined the IQs of children living in institutions in Russia, the majority of whom were special needs children. The researchers implemented two different interventions: one intervention involved training to educate staff on aspects of early childhood, emphasizing warm, caring, sensitive, responsive, and developmentally appropriate interactions. A second intervention, implemented along with staff training, involved making structural changes to the context of care, mainly reducing group size from approximately twelve to six, assigning two primary caregivers to each group so that a primary caregiver was available every day, reducing transitions of children to new wards or caregivers, and establishing periods of time when caregivers were instructed to be with the children. The study also observed children in a third institution where no changes were made. Thus staff at the institutions could get training alone or training plus structural change or no intervention. The authors found that the training plus structural change group showed marked and significant improvements in IQ compared with the training alone group as well as with the no-intervention group. Modifying the physical context and training caregivers in that context affected IQ in a positive manner among institutionalized infants. However, again, IQ scores of children in the most intensive intervention group did not match those of a community control sample. In general, changing the environment in an institution has some positive effects for IQ, though multiple factors, such as age of entry into the institution and duration of institutionalization, have not been adequately investigated.

IQ of Institutionalized Children Adopted or Reunited within Country

Studies of institutionalized children who were either adopted within their country of origin (so-called intracountry adoption) or reunited with their biological families find mixed results with regard to IQ. Barbara Tizard followed institutionalized children in the UK in these different placements, comparing them with children who remained in institutions as well as with a community sample.[27]

At four years of age, twenty-four of the sixty-five children originally enrolled in the study had been placed in adoptive homes and fifteen had been returned to their biological mothers.[28] Children adopted at a young age had scores on the Wechsler Intelligence test (WPPSI; see Chapter 7) that were approximately one standard deviation (15 points) higher than those of similarly aged children reunited with their biological mothers. Despite these differences, both scores were within the average range, suggesting no serious cognitive deficits for these youngsters.[29] By the age of eight, children adopted before 4.5 years of age scored well above the average in both IQ and reading attainment of children placed later.[30] This pattern held when the children were re-examined at age sixteen.[31] Institutional rearing did not have the devastating long-term effects described in some early studies, as evidenced by children's normative intelligence (mean IQ at age sixteen of children adopted before four years: 114; mean IQ at age sixteen of children adopted after four years: 102).[32] Notably, however, these institutional environments had smaller child-group sizes, better child-to-caregiver ratios, and were materially better supplied than the Romanian institutions studied in the BEIP.

Moreover, different placements were associated with different IQ gains. The largest IQ gains were apparent in children placed in adoptive homes between the ages of two and 4.5 years, gains that were maintained over the following twelve years. It is important to note that children were not randomly assigned to adoption or to reunification with their biological families, and these results must be understood in that context. These findings are somewhat at odds with the meta-analytic conclusions reviewed above. They no doubt reflect the backgrounds of the children and the quality of the institutional environment.

A study in Greece, focused on sixty-two children who had been adopted after spending the first year to year and a half of life in an institution, found less positive effects.[33] The children were followed up when they were four years of age and compared with a typically developing same-age control group. Results showed that the adopted children had lower IQ scores compared with the control group (mean of the adopted group: 100, mean of the controls: 110). It is interesting to note that the IQ scores of the adopted children were related to their IQ scores in infancy, suggesting that, unlike in Tizard, where previously institutionalized children at an older age appear to catch up in IQ, the children in this study did not have a long enough period of time to "catch up" in cognitive growth.

IQ of Post-Institutionalized Internationally Adopted Children

International adoption has been a major vehicle by which parents, mostly in the West, have been able to adopt young children. However, much controversy has surrounded international adoption, with stories of child-trafficking flooding the Internet. Not surprisingly, politics has weighed heavily in this avenue for finding permanent placements for parentless children. For example, ten years ago more than 20,000 children were adopted internationally into homes in the United States, whereas in 2012, this number approached 10,000.

Over the past twenty years, a number of research teams have taken advantage of the rise in international adoption.[34] Virtually all studies of internationally adopted children show that these children display deficits in IQ when they enter their adoptive homes. Most of these studies, however, also find catch-up in IQ for these children. One such study compared young girls adopted from China with children adopted from foster care in the Netherlands.[35] The researchers found that the Chinese children at first showed significant deficits, but that over time they caught up to their peers from foster care. Indeed, in a review of these studies, Marinus van IJzendoorn and Femmie Juffer found that children placed into families catch up in IQ compared with their peers who remain in institutions.[36]

In the case of Romania post-Ceauşescu, two longitudinal studies, one by Elinor Ames and colleagues and the other by Michael Rutter and colleagues, followed children after adoption from institutions. The Ames study consisted of three groups: forty-six children who were institutionalized for at least eight months; forty-six non-adopted, never-institutionalized Canadian-born children; and twenty-nine children adopted early (before four months of age) from Romania. At time of adoption, all children older than four months were found to be developmentally delayed. At follow-up (three years post-adoption), these same children continued to have significantly lower IQ scores compared with the other two groups.

Children adopted from institutions after two years of age had the lowest IQs, with an average score of 69, and 56 percent of these children were deemed not ready for the school level appropriate to their age, compared with 17 percent of children adopted out of institutions before four months of age. In the later-adopted children, IQ scores were lower the longer the child spent in the institution.[37] These results speak to the especially harmful effects of institutional environments in Romania in the 1990s and also to the importance of early adoption.

A similar pattern of findings has been reported in the ERA, which examined institutionalized Romanian children adopted into UK homes before forty-two months of age and followed to eleven years of age.[38] Using a comparison group of children born and adopted within the UK before six months of age, they found that Romanian adopted children were significantly impaired in their cognitive development at the time of entry into the UK. When examined during a follow-up at age four, Romanian children who were adopted before the age of six months showed significant gains, whereas those children adopted after six months were still delayed in their cognitive development.[39] By age six, children adopted before six months of age no longer differed from the community group in their cognitive scores, and children from both of these groups had scores that were significantly higher than those of children adopted after six months of age. Similar results were found at age eleven; however, only those children who were in the late-adopted group experienced significant increases in their cognitive scores between the six- and eleven-year assessments, potentially indicating a late-occurring period of catch-up for these children.[40]

These impressive studies have contributed a great deal of invaluable data about the cognitive effects of early institutional rearing.

The challenge to interpretation with both the Ames study and the ERAS is the possible bias in selection of the children being adopted and the lack of detail regarding the quality of early experiences of these children. In other words, it is possible that the children who were fortunate enough to be adopted were somehow different from those left behind: they could have been healthier, better-looking, more socially engaged, and so on. A case in point: in the Tizard studies mentioned earlier, red-headed children were less likely to be adopted than non–red-headed children.[41]

In sum, conditions of severe psychosocial deprivation such as those that existed in the institutions in Romania had significant effects on IQ development in young children. Data suggest that while very early limited exposure to such conditions does not have long-term effects, exposure continuing beyond six months of age leads to long-term decrements in intelligence. Variability across studies appears to be a result of the types of caregiving environments of institutions in different countries as well as the age at which children enter the institution and the duration of their stay.

Executive Function

Although IQ is viewed as a measure of general intelligence, psychologists understand that it does not provide the specificity of cognitive processes that are important in problem solving and, for that matter, in daily functioning. In other words, IQ really reflects an amalgam of different cognitive processes. Executive function processes, on the other hand, including inhibitory control, working memory, and attention shifting, are well established as being involved in adaptive functioning.[42]

Seth Pollak and colleagues assessed memory, attention, and executive function in post-institutionalized children using two well-validated test batteries, the Cambridge neuropsychological test automated battery (CANTAB, a computerized series of neuropsychological tests) and the NEPSY (Neuropsychological Assessment) developmental neuropsychological assessment. Internationally adopted children between the ages of eight and ten with a history of prolonged institutionalization performed worse on tests of memory, visual attention, and learning than did early-adopted or non-adopted control children. More time spent in an institution was associated with poorer performance on tests of inhibitory control, visual attention, and visual memory/learning. These findings clarify some of the specific neurodevelopmental sequelae of early deprivation and support the hypothesis that children who experience significant periods of early institutional care show impairment patterns in important cognitive processes associated with problem solving and complex social cognition.[43]

Further exploring the association between early deprivation, executive function, and social cognition, Rutter and his colleagues examined the link between institutional deprivation, executive function, and theory of mind in the ERA sample.[44] Romanian adoptees displayed deficits on executive function measures as well as in theory of mind, with the effects strongest among those who had experienced more than six months of institutional deprivation. As a further link to the proposed institutional syndrome, deficits in both domains were associated with quasi-autism, disinhibited attachment, and inattention/overactivity.

In sum, the few studies that have examined specific cognitive processes including executive function in institutionalized children have found deficits in these samples. Such deficits have significant consequences for everyday functioning and for the development of social cognitive processes such as theory of mind as well as basic processes of attention.

Language Development

Deficits have been noted in language development in both children living in institutions and children adopted out of institutions.[45] Institutionalized infants vocalize less and have fewer types of vocalizations as early as two to four months of age when compared with infants raised in their own families.[46] Lack of imitation by caregivers and interaction with them, coupled with likely reductions in exposure to mature language, were cited as contributors to this early differentiation in language skills. Foster children who were institutionalized for the first three years of life showed more speech problems than children raised in foster homes.[47]

The importance of language as a means of interaction and also of supporting the development of basic cognitive skills makes delays or deficits in this area especially salient.[48] Language is also an important factor in the child's development of self-regulatory skills. Children develop language in environments characterized by both stimulation and contingency.[49] It is not surprising, therefore, that in studies of children adopted into the United States from Eastern Europe, speech and language delay is one of the primary deficits.[50] The foundations of language skills, such as turn-taking, develop long before the child speaks his or her first word. Contingent responsiveness on the part of the caregiver is an integral part of this process. When the child vocalizes and the caregiver responds, the child becomes a partner in the language experience. In an institutional setting, harried caregivers who are responsible for changing a room full of children cannot make the time to respond to the communication attempts of each child. As a result, the infant eventually stops trying to communicate.[51] Institutionalized and post-institutionalized children routinely have been shown to have difficulty with speech and language skills, more so for expressive than for receptive skills, but delayed nonetheless. In several studies, rates of language delay approached or exceeded 70 percent, and these rates persisted even when children had been placed in foster care as preschoolers.[52]

Surprisingly, not many recent studies have formally examined the language of institutionalized children. In part, this is because children adopted internationally learn the language of their newly adopted home, and their cognitive abilities are intricately linked to their language skills. There also appears to be a good deal of plasticity in language learning in young children, hence their ability to learn the new language of their adopted family. Yet the circuitry and brain regions that support language are intimately involved in cognition, and so the very same effects that early exposure to deprivation has on cognition may be affecting language as well.

Attachment and Social/Emotional Development

Of all the domains studied among previously institutionalized children, the one in which children show the greatest deficits is social and emotional behavior. This deficit is particularly pronounced in attachment relationships, most likely because of the significant absence of human social interaction within institutions across the early years of life and the need for such input for adequate connectivity in those brain regions associated with typical social behavior.

Human infants are born into a world in which the experience-expectant environment involves—or at least should involve—sensitive and responsive caregiving and the availability of contingent human social interaction. The context of the institution, therefore, presents a significant problem with regard to the availability of adequate experiences for adaptive social development. Over the first year of life, most infants exposed to typical experience-expectant social contexts develop relationships with a number of adults who care for them and furnish a protective, loving interactive environment. An attachment relationship between a young child and a caregiver requires sufficient ongoing interaction between the two so that the child learns to seek comfort, support, and nurturance selectively from that caregiver. Because this opportunity is limited in institutions, we would expect children raised in such settings to be unable to form attachments. Across the many studies of children in institutions, a number of themes recur with regard to attachment and social relationships. These include observations of children as being unable to form deep and intimate relationships with their adopted parents and behavior that involves indiscriminate engagement with adults. These patterns are observed while children live in the institutions and even after they leave them—sometimes many years after a child has been living with adopted parents in a stable family environment.

Initial studies of institutionalized children reported social behaviors that clustered around two different dimensions: externalizing behaviors (such as inappropriate approach and engagement with adults) and internalizing behaviors (such as social withdrawal and anxiety). Some reports noted that children who experienced severe maternal deprivation were unable to form deep and intimate relationships, even after they were adopted into warm family environments.

The quality of the attachment between a caregiver and an infant is gauged by the manner in which the child organizes his or her responses to caregiver separation and reunion. Typically research studies use a procedure called the Strange Situation (see Chapter 10), which is a series of separations and reunions between infant and caregiver. Children organize their responses to separation and reunion differently depending upon their experiences with a caregiver. Organized secure responses involve seeking proximity with the caregiver at reunion and being comforted by the caregiver after separation. Organized insecure responses involve either avoiding the caregiver at reunion or displaying distress or anger at the caregiver and not being easily comforted. In addition, infants might display what has been called a disorganized response to separation and reunion. This typically involves a breakdown of attachment strategies (of proximity seeking) and undirected or misdirected movements, freezing or aimless wandering. It is important to note that typically infants mostly form organized attachments whether secure or insecure. Disorganized attachment is more likely in populations of infants who have experienced adversity, including infants who have experienced extreme psychosocial deprivation.

The early studies of Goldfarb and Provence and Lipton describe the pattern of shallow emotional responses of children in institutions toward caregivers: anxiety, withdrawal, and "over-friendliness."[53] The studies of Tizard and Rees found that among twenty-six continually institutionalized children, most had seriously disturbed attachments. Ten of these children were described as "overly friendly," that is, they approached unfamiliar adults as often as familiar adults. Evidence of reticence around unfamiliar adults was absent, and these children sometimes exhibited separation protest when leaving strangers. They displayed attention-seeking behavior and were indiscriminate in seeking comfort. The children's attachment to others seemed superficial since for them adults appeared interchangeable.[54]

In a sample of infants raised in a Greek institution, Panyiotta Vorria and colleagues reported that a majority of these infants (65 percent) had "disorganized" attachments to their caregivers.[55] Juffer and colleagues used a standardized procedure to assess attachment in internationally adopted children in the Netherlands in infancy. Most infants (74 percent) were securely attached at twelve months, whereas 22 percent were disorganized, and the sample did not deviate from normative groups on attachment classification. These early-adopted children (mean age at arrival was eleven weeks) came from relatively favorable backgrounds: the Sri Lankan children were cared for by their birth mothers until placement, and the Korean and Colombian infants came from children's homes supported by Western agencies.[56]

Three studies have examined patterns of attachment in preschool children adopted out of Romanian institutions. These studies all have found a high proportion of unusual classifications. Sharon Marcovitch and colleagues found that only 30 percent of young children adopted from Romania were securely attached (compared with 42 percent in a Canadian-born never-institutionalized comparison group), and 42 percent were insecure/controlling (compared with 10 percent in the comparison group). Kim Chisholm reported more insecure attachments in Romanian adoptees (63 percent) than in Canadian-born comparison children (42 percent). Additionally, more than 21 percent of the Romanian adoptees displayed atypical insecure patterns, while none of the Canadian children did so. Finally, Thomas O'Connor and colleagues found that at age six, following adoption into the UK out of Romanian institutions, 51 percent of children were classified as insecure/controlling or insecure/other compared with only 17 percent of children adopted within the UK (but never institutionalized). In addition, 36 percent of children adopted out of Romanian institutions, compared with only 13 percent of children adopted within the UK, exhibited non-normative behavior in separations and reunions, such as extreme forms of emotional over-exuberance, nervous excitement, silliness, coyness, and excessive playfulness more typical of much younger children.[57]

Psychopathology

Much like the research on development of attachment relationships to caregivers, studies of post-institutionalized children's psychopathology have identified similar deficits. The most consistent finding pertains to inattention/hyperactivity.[58] William Goldfarb was first to suggest that children who experienced institutional care and were removed to foster care at around age thirty-six months were more likely to experience externalizing behaviors such as aggression, hyperactivity, and destructiveness; they were also more likely to have difficulty appreciating the feelings and needs of others and forming attachments. Children who had been institutionalized for the first three years of life demonstrated consistently greater levels of problem behaviors, ranging from feeding and sleeping difficulties to aggressive behavior and hyperactivity to marked over-dependence on adults characterized by attention-seeking behavior.[59]

Michael Rutter and his research team (the English and Romanian Adoptees study group) have also identified problems in attention and hyperactivity among the children they studied. In particular, children adopted into UK homes after the age of six months display what Rutter has called an institutional deprivation syndrome that includes inattention and hyperactivity, cognitive impairment, indiscriminate friendliness, low IQ, and quasi-autistic behaviors. Descriptions of children from the ERA affirm reports of restlessness, hyperactivity, and difficulty with attention, problems that have been shown to be reasonably stable patterns of behavior with long-term implications.[60] Rutter argues that this constellation of behaviors is not seen in other circumstances and appears to be the result of early psychosocial deprivation. For example, Rutter has shown that patterns of hyperactivity, as well as difficulty with attention among previously institutionalized children, are linked with difficulty in establishing and maintaining selective attachments.[61] Again, this institutional syndrome is somewhat similar to the constellation of behaviors described many years earlier by Goldfarb. He reported that at age six, children who were raised in institutions for the first three years of their lives displayed more hyperactivity, anxiety, and difficulty with attention than did their counterparts raised exclusively in foster care, as reported by social workers familiar with them.[62] Similarly, Lowrey noted a pattern of hyperactivity, particularly as formerly institutionalized children were transferred into "boarding" or foster homes. Tizard and Rees reported that adopted children were less distractible and restless than children either remaining in the institution or reunited with their families.[63]

Aside from the prominence of attention problems and hyperactivity and their link to indiscriminate social behavior, Rutter also reported that autistic-like behavior was part of an institutionalization syndrome. Rutter and colleagues found that 6 percent of the ERA sample met diagnostic criteria for autism and that another 6 percent had mild, often isolated, features of the disorder. Given that most children were placed in institutions at or shortly after birth, and given the proportion of children affected, it seemed not that this reflected a placement bias but rather that the experience of institutionalization produced a phenotype similar in nature to autism. Interestingly, when re-evaluated at age six, most of the children in this sample were much improved and no longer met criteria for autism, though some continued to exhibit unusual behaviors. Because the children had responded so dramatically to more favorable environments, Rutter and colleagues described the syndrome as "quasi-autistic." The vulnerabilities in individual children that might lead to such a clinical picture among institutionalized children and the processes through which the caregiving environment ameliorates such symptomatology remain to be elucidated.[64]

Neuroimaging Findings among Institutionalized Children

One approach to understanding the mechanisms by which institutionalization affects behavior is through an examination of the neural correlates of cognitive and social functioning among children who have experienced institutionalization. The BEIP was the first study to assess brain functioning in institutionalized young children. Nevertheless, some studies before and after BEIP have examined brain structure and functioning in formerly institutionalized children. We review these briefly in this section.

A few recent studies have used brain-imaging methods to examine the differential use of glucose by different brain regions (called Positron Emission Tomography, or PET imaging), the size and integrity of brain structures (called structural imaging), the integrity and distribution of fibers that connect different areas of the brain (called Diffusion Tensor Imaging, or DTI), and the response of different regions in the brain to tasks examining cognition or emotion (called functional imaging). These last three tests use MRI methods.

PET is a radiographic method that involves injecting an individual with a substance like glucose or oxygen that has been radio-labeled. The radioactive substance is then absorbed by the brain, with those brain regions most engaged (most active) taking up more of the substance. As the radioactivity begins to decay, positrons are emitted, and the PET scanner can reconstruct where in the brain these positrons originated. As such, it is possible to pinpoint which precise regions of the brain are most active. Problems with this method include the need for radioactive substances being injected into the bloodstream that are then absorbed by different brain regions. By contrast, MRI is a noninvasive method in which a magnetic field, along with pulsed radio waves, is used to alter the alignment of atoms in the body. These changes in alignment are then detected and images are created displaying the underlying structures in the body (think of the MRI one gets of a knee or shoulder). Innovative methods have made it possible to image the structures of the brain using this approach. The structures that can be imaged include what is called gray matter (such as the neurons of the cortex) and white matter (the myelinated axonal tracts that connect different parts of the brain). In addition, with subjects performing tasks while lying in the MRI scanner, clinicians can image changes in blood flow to different parts of the brain, hence examining functional changes in brain activity.

Any of these studies is challenging to complete, requiring sophisticated and expensive equipment and data-processing methods. Subjects for studies on institutions often come from different countries (for example, China, Russia, Romania), with variations in ethnicity, gender, and length of experience in institutions. Nevertheless, these studies are informative in attempting to identify structures involved in cognitive and social behavior that are compromised either structurally or functionally.

In one of the first studies to examine the effects of early institutionalization on brain development, Harry Chugani and colleagues used PET scans to study ten children who had been adopted from a Romanian institution. In the original study, the average age at study was eight years, and nearly all children had been placed in the institution before eighteen months and had lived there an average of thirty-eight months before being adopted. These children were compared with a group of healthy adults and a group of ten-year-old children with medically refractory epilepsy. The authors reported that the adoptees showed significantly reduced brain metabolism in select regions of the prefrontal cortex (for example, orbitofrontal cortex) and the temporal lobe (for example, the amygdala). These, of course, are regions typically associated with higher cognitive functions, memory, and emotion. Neuropsychological testing revealed that the adopted children suffered from mild neurocognitive impairments, including impulsivity, attention, and social deficits. Because these were the first imaging data to be reported on post-institutionalized children, they received a great deal of attention in the popular press, for good reason: the findings were dramatic. However, one must be cautious in making too much of these findings, given that the sample size was small and highly selected (both in who was allowed to leave the institution and in who consented to have a radioactive isotope injected into the bloodstream).[65]

This same group of investigators conducted a follow-up study of this sample of children, this time examining white matter connectivity using MRI.[66] They found that white matter connectivity was diminished in the uncinate fasciculus region of the brain in the early deprivation group compared with controls. The uncinate fasciculus provides a major pathway of communication between brain areas involved in higher cognitive and emotional function (for example, amygdala and frontal lobe), leading the authors to conclude that connectivity between brain regions is significantly reduced by early institutionalization. Fewer connections may lead to deficits/delays in inhibitory control, emotion regulation, and other functions that are overseen by the connections between these brain areas.

In a third study by Chugani and colleagues, again using MRI, a highly sensitive neuroimaging analysis technique, Tract Based Spatial Statistics (TBSS), was employed to evaluate further white matter changes in children with a history of early deprivation. Seventeen children with a history of institutional care since birth and subsequent adoption were compared with fifteen typically developing non-adopted children. The children came from a diverse range of institutions, including in Eastern Europe and Southeast Asia. They were found to have reduced organization of portions of the bilateral uncinate and superior longitudinal fasiculi, with the magnitude of white matter disorganization significantly associated with duration of time spent in institutional care. The authors speculate that these findings may underlie the inattention/overactivity syndrome that was previously described by Rutter and colleagues as part of the institutional syndrome. Collectively, work by Chugani and colleagues suggests that children who experienced early institutionalization suffer from metabolic, structural, and connectivity deficits in the areas of the brain believed to be involved in higher cognition, emotion, and emotion regulation.[67]

Three studies to date have used MRI methods to examine children who were living in the United States after being adopted from institutions around the world. P. M. Bauer and colleagues completed a structural imaging study of thirty-one post-institutionalized children living with adopted families in the Midwest. The children were mainly from institutions in Eastern Europe (twelve from Romania, twelve from Russia, five from China, and two from other European countries). Their average age at adoption was thirty-one months, and they were tested at a mean age of eleven years. The post-institutionalized children had smaller volumes of the superior-posterior cerebellar lobes, and poorer performance on memory and executive function tasks compared with typically developing children who served as controls. Structural differences in cerebellum mediated the relation between institution history and cognitive performance. This is one of the few studies to identify the cerebellum as being affected by early experience and related to higher-order cognitive function. The cerebellum is a structure deep in the brain that is involved in motor control and basic learning processes. The fact that early experience of deprivation can affect this basic structure helps account for the deficits seen in post-institutionalized children in attention and cognition.

Nim Tottenham and colleagues published two imaging studies with post-institutionalized children from Asia and Eastern Europe, examining structural differences between these children and a typically developing control group.[68] Findings from the first study showed no differences in cortical volume or gray matter between the different groups of children, though late-adopted children had larger amygdala volume compared with the early-adopted and the typically developing controls.[69] In their second study, they report finding greater amygdala activity in the post-institutionalized group in response to the fear stimuli, with amygdala activity mediating the relation between early institutional care and eye-gaze pattern (post-

institutionalized children showed less direct eye fixation on faces).[70]
Coupled with the first study's finding of enlarged amygdala in post-
institutionalized children, these studies showed the importance of
this subcortical structure and its apparent sensitivity to deprivation
experiences in children. Importantly, in both studies, the imaging
took place years after the children had left the institution and were
now adopted into stable families in the United States, thus emphasizing the effects of early experience on amygdala volume and activity.

What is the significance of these amygdala findings? The amygdala is an almond-shaped structure deep in the middle of the brain
that is involved in the detection of threat and unfamiliar events or
people. It helps in the processing of emotional information from
others' faces, and particularly fear learning. The amygdala plays a
role in the processing of emotionally salient information both in
terms of the formation of emotional memories and the guiding of
behavior based on emotional/threat-related stimuli through the
modulation of attention of other areas of the cortex.[71] As a basic
structure in the brain involved in fundamental responses to stimuli
(novelty detection, approach-withdrawal), it plays a significant role
in many social and cognitive behaviors. Hence, disturbances in
amygdala activity and structure as a function of early experience
have a significant impact on more mature and complex social and
cognitive processes.

A recent paper from the English Romanian Adoptees study used
MRI to examine amygdala structure among post-institutionalized
children. The study found that the adoptees had significantly reduced total gray and white matter volumes compared with the
control group. After correcting for brain volume, the adoptees had
enlarged relative amygdala volumes compared with the control

group, particularly in the right hemisphere. Left amygdala volume correlated significantly with the duration of institutional care, with longer stays in an institution correlated with a smaller left amygdala volume.[72]

As can be seen from the brief overview of brain-imaging studies of post-institutionalized children, the findings are highly variable: some find increased amygdala size along with reduced gray and white matter volume, while others find association between structural size or function and duration of institutionalization. The variability in findings is most likely a function of the heterogeneity of experiences, contexts, and durations of exposure to deprivation across the different samples, as well as the small sample sizes in any particular study. Nevertheless, the findings provide important guideposts for thinking about the effects of early experience on brain development and understanding the mechanisms by which early experience influences cognitive and social behavior.

Chapter 7

Cognition and Language

The multiplicity of individuals that share in caring for institutionalized infants results in a fragmentation of care and lack
of constancy that are believed to make it more difficult for
an infant to develop an awareness of himself and his environment. It appears to exert a retarding influence on learning in
a general sense.

Sally Provence et al., *Infants in Institutions* (1967), pp. 18–19

Both clinical observations and the scientific literature point to
the fact that children who are placed in institutions early in life
show declines in their cognitive abilities as well as in their language
function. Such declines vary as a function of time spent in an institution, with greater declines the longer a child is in institutional care.

In the BEIP, we were interested in evaluating children's cognitive and linguistic abilities prior to randomization and continuing
through all subsequent follow-ups, using a combination of standardized measures of intellectual function (for example, developmental and IQ tests), as well as experimental neuropsychological
measures that permitted us to draw inferences about specific cognitive functions (for example, inhibitory control) and their relation to
underlying brain circuitry (see Table 7.1 for a list of the measures
we employed).

Table 7.1 Measures used to evaluate cognition, language, and neuropsychological function

Construct	Measures	Assessment schedule					Source of data
		Baseline	30 mos.	42 mos.	54 mos.	8 yrs.	
Cognition	Bayley Scales of Infant Development	X	X	X			exam of child
	Wechsler Preschool and Primary Scales of Intelligence				X		exam of child
	WISC–IV					X	exam of child
Language	REEL		X	X			exam of child
	Reynell		X	X			exam of child
	Peer Interaction Tasks			X		X	transcript of child speech
	Caregiver Interaction Tasks			X			transcript of child speech
Neuropsychological function	Bear-Dragon Task				X		exam of child
	Posner Spatial Cueing					X	exam of child
	Go-No-Go					X	exam of child
	Mismatch negativity	X	X	X			exam of child
	Flanker					X	exam of child
	CANTAB					X	exam of child

Intelligence Quotient

IQ is a measure of intelligence that has for many years been used as a standard by which to judge differences among groups of individuals. What we think of now as the standard IQ test was first developed by Alfred Binet, who, along with Theodore Simon, published the Binet-Simon test for determining a child's mental age. Louis Terman, a psychology professor at Stanford University, revised this test (now called the Stanford-Binet Test of Intelligence), and the Intelligence Quotient (IQ) score became the most widely used test for assessing child and adult intelligence.[1]

David Wechsler in 1939 devised another version of the IQ test, one that examined multiple aspects of intelligence, such as spatial ability and verbal comprehension (later known as the Weschler Intelligence Scale). There are now multiple versions of the Wechsler Intelligence test, including one for preschool children (WPPSI), one for older children (WISC), and one for adults (WAIS).

Generally, IQ tests are normed on large populations of individuals; the mean IQ score for a population is 100 with a standard deviation of 15. This means that the majority of people (70 percent) score between 85 and 115 on the IQ test. IQ tests such as the Wechsler versions are normed down to age three.

Below age three, standardized measures are used to assess infant development. One of the most popular measures was developed by the psychologist Nancy Bayley. The Bayley Scales of Infant Development provide a measure of infant development called the Mental Development Index (MDI), which is a scaled score, ranging from 50 to 150. In BEIP, children who obtained scaled scores below 50 were assigned an MDI score of 49. Raw scores were assigned an age-equivalent score to enable analyses when scaled scores below 50 were obtained. Developmental Quotients (DQ) were computed for each child (that is, [age equivalent score/chronological age] x 100 = DQ), allowing analyses to be carried out on the entire sample.

Before randomization, we administered the Bayley Scales of Mental Development to the children living in the institutions. In follow-up visits, after randomization and placement of half of the sample into our foster care intervention, we continued to use the Bayley Scales because the Bayley is normed through forty-eight months.[2] We switched to the Wechsler preschool version of the IQ test at the fifty-four-month session and to the WISC-IV when children were tested at age eight.[3] Research on IQ scores has found stability of scores across ages and types of tests for typically developing populations.

Examining the scores of the children at baseline revealed that the institutionalized children had an average MDI of 66 while community children had an MDI of 102.[4] The average Developmental Quotient for the institutionalized children was 74, whereas it was 103 for community children. This is almost two standard deviations below the mean for same-aged children. At baseline, the institutionalized children ranged in age from six to thirty months, and there was a negative correlation between age and MDI. That is, the older the child (and therefore the longer the child had been institutionalized) the lower that child's DQ score was at baseline. The scores of the institutionalized children at baseline suggest profound intellectual delay.

We also examined differences in caregiving quality across the institutions in which the children had been living. We used a standardized observational measure called the Observational Record of the Caregiving Environment (ORCE), which had been used previously in the National Institute of Child Health and Human Development (NICHD) study of child care.[5] The ORCE is composed of behavioral scales, in which occurrences of specific caregiving acts with the child are recorded, as well as *qualitative ratings,* in which the quality of the caregiver's behavior in relation to the child is evaluated. Some of the features captured in the qualitative ratings include shared positive affect, positive physical contact, and positive response to vocalization/child's talk. We found that the better the caregiving in a particular institution the higher the IQ of the child. Nevertheless, the very low scores of the children in the institution underscored the magnitude of the effects of severe deprivation on children's IQ even at this young age.

Immediately after the baseline assessments, the children were randomized into the two arms of the study. Thus children varied in their chronological age at assignment. Some children were as young as six months and others had already had their second birthday. The next three assessment points occurred when all the children were thirty, forty-two, and fifty-four months of age. This meant, for example, that for the thirty-month assessment, some children who went into foster care at the age of twenty-eight months might have had only two months of the intervention while others who were eight months at placement would have had twenty-two months (see Figure 2.3). The Bayley II was administered at baseline, thirty, and forty-two months of age.[6]

We asked two questions of the IQ data collected at these three age points of thirty, forty-two, and fifty-four months. First, was there an effect of the intervention? That is, did the children randomized to the foster care group outperform the children randomized to the institution? Second, within the foster care group was there an effect of age of placement or duration of intervention on IQ?

Inspection of the mean IQ scores for the foster care and institutional groups at each age indicated that the children placed into foster care outperformed their peers who had remained in the institution. Although this was very good news and demonstrated the effectiveness of our intervention at each age, the children placed into foster care still lagged significantly behind typically developing community children from Bucharest. This may speak to the prenatal risks to which the institutionalized children were exposed, the familial risk factors they carried, the "relatively" late age of placement, or limits in the intervention itself.

We next examined the relation between age of placement and IQ at each age, as well as length of time in intervention and IQ at each age for the children placed into foster care. We found that only age of placement and not duration of time in foster care was positively related to IQ scores at each age (see Figure 7.1). That is, age upon leaving the institution and entering foster care was more important than time spent in the foster care home. Of course, both these factors—age at placement and length of time in care—are highly correlated with each other. Nonetheless, when we examined which was more powerful a predictor of IQ, the answer was age at placement. Indeed, within the confines of the distribution of ages in our sample (ages of the children at placement), we identified a cut-off point for age of placement before which children placed into foster care outperformed their institutionalized peers, and after which there was little difference between a foster care child and a child randomized to remain in the institution.

The age that appeared to divide the children in terms of per-

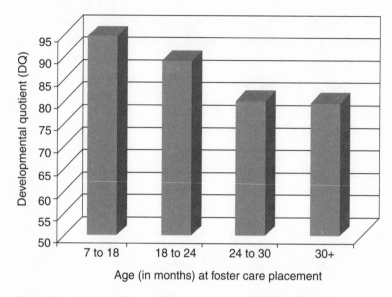

Figure 7.1. Mean Developmental Quotient (DQ) for the children randomized to the foster care group as a function of their age of placement into foster care. Note that DQ, a measure derived from the Bayley Scales of Infant Development, is analogous to an IQ score.

formance was twenty-four months. If a child was younger than twenty-four months at placement, then at each of the three ages, thirty, forty-two, and fifty-four months, he or she was more likely to outperform children older than twenty-four months at placement. This finding strongly suggests a sensitive period in cognitive recovery, such that children placed after the age of twenty-four months more closely resembled institutionalized children and those placed before this age began to approximate never-institutionalized children. That said, this twenty-four-month cut-off point is no doubt in part a function of the ages of children included in our sample, as well as how the children's ages were distributed. For example, Michael Rutter and the ERA group found that those children who were adopted from Romania below the age of six months recovered their intellectual function. That is, there were no differences between this group and controls. We found an intervention effect and timing effect at fifty-four months for those children placed into foster care before twenty-four months of age. But even though those children placed into foster care below twenty-four months as a group performed better than those placed after twenty-four months and those randomized to the institution, their mean level IQ scores were still not comparable to those of typical controls.[7] The next assessment point for the children in the study was when they were eight years old.[8] At that time, we administered the WISC-IV, which provided us with scores for each child in full-scale IQ (a compilation of multiple subscales that each assesses a different element of intelligence, such as short-term memory), as well as four subscales (processing speed, verbal comprehension, perceptual reasoning, and working memory).

Our first approach was the "intent to treat" analysis in which children who were originally assigned to either the foster care or the care-as-usual groups were analyzed as if they continued to live or remain in those caregiving settings. Although this approach to analyzing the data was inherently conservative, it had the virtue of providing a strong test of our hypothesis that early experience contributed more to child outcomes than did subsequent life experience. When using this approach we found no differences on any of the subscales or on full-scale IQ between the two groups. The full-scale IQ of the foster care group was 81.5 and for the CAUG it was 76.2. It appears that while the IQ scores of the foster care children remained stable over time, the scores of the children randomized to care as usual began to catch up. Note that, as before, the scores of both groups were far below the mean scores of the community controls. In addition, the timing effects we had identified at thirty, forty-two, and fifty-four months were no longer evident.

What happened to the IQ differences between the foster care and the institutionalized children found earlier? There are several plausible explanations. First, the initial boost in IQ that we observed in early childhood (through fifty-four months) might be an immediate but unsustainable effect of being removed from the institution and placed into families. Second, the lack of group differences at age eight might be a function of changes in schooling context for the children in the study. That is, both the foster care and the care-as-usual children started school at age seven, and hence the increase in the care-as-usual mean IQ levels could compensate for the lack of the earlier IQ boost in the foster care sample. Third, the lack of differences at age eight may portend sleeper effects (that is, differences that may emerge later as a function of the early intervention but that do not appear in early childhood). Such a pattern is not uncommon in intervention studies with children in the United States of low socioeconomic standing (SES).[9] Fourth, we can address the differences in results between the early IQ data and the eight-year data by examining Figure 2.2, which illustrates the change in group assignment over the first eight years of the project. This figure shows that only fourteen of the original sixty-eight children randomized to remain in the institutions still lived in institutions. The remainder had been reunited with their biological parents, placed into government foster care (not available when the study began), or adopted by Romanian families. Similarly, of the sixty-eight children randomized to our foster care intervention, thirty-one were still in these homes, whereas the other children had been either adopted or reunited with their biological families.

So there was a great deal of movement in the lives of the children in our study between four and a half and eight years of age.

We decided to investigate further the pattern of IQ results by examining scores of children based on their current placement status. In other words, instead of using intent to treat analysis, we examined the effect of current living context on IQ at age eight. We wondered whether children who were originally assigned to BEIP foster homes and remained there would do better than children who stayed in the institution or were placed into government foster care. The answer was a resounding yes. Children who remained in BEIP foster homes had higher scores than the two other groups on three of the four subscales (verbal comprehension, processing speed, and working memory) and higher full-scale IQs. As a further check, we found that among the foster care group those who had remained in the intervention homes did better than those who had left. These two sets of results clearly show the benefit of a stable, positive family environment on IQ scores. While the initial uptick in IQ scores was due to our intervention, sustaining those intervention effects was a result of the child's remaining in a stable placement over time.

We then decided to make use of the longitudinal nature of our data, examining the pattern of IQ scores in both the FCG and the CAUG groups. In each group there were two obvious trajectories over time. One involved children who had stable but low scores across age, while the other involved higher stable scores over time. We examined predictors of these different trajectories and found for the FCG group that attachment security and positive, responsive caregiving at forty-two months predicted a high and stable pattern of IQ. In addition, strong language skills at forty-two months contributed to these stable high patterns. We could not, however, find predictors of patterns within the CAUG group, except for poor language abilities at forty-two months related to low and stable trajectories in IQ.

One more interesting finding emerged from these data. Examining the trajectory of children who continued to live in the institutions these eight years, we found a progressive decrease in IQ with age. To our knowledge, this is the first ever longitudinal study of IQ scores on children living in institutions, and it shows a decline in IQ over time. This underscores the effect of institutional life on children's intellectual functioning and again supports the fact that the longer a child remains in the institution the lower his or her IQ is likely to be.

Neuropsychological Functioning at Eight Years

IQ is an invaluable index of a child's overall intellectual ability, and it correlates reasonably well with academic success. However, since the BEIP was conceptualized as a project designed to explore how brain development is influenced by experience, we felt it essential to include measures of cognitive function that would permit reasonable inferences to be drawn about underlying brain architecture. We were particularly interested in tasks that tapped the functioning of the prefrontal cortex (PFC), as the PFC is the primary seat of executive functions (EFs). EFs represent a collection of skills and abilities, including working memory, planning, cognitive flexibility, inhibitory control, and executive attention.[10]

Studies with nonhuman primates suggest that the prefrontal cortex and its associated systems may be especially vulnerable to early experience and early adversity.[11] Moreover, because the PFC undergoes protracted development that may not be complete until mid-to-late adolescence, this region may remain particularly sensitive to experience-dependent fine-tuning.[12] Of course, in the case of institutionalized children, such fine-tuning may have negative consequences for development. Indeed, on the basis of reports from others studying institutionalized children (see Chapter 6) as well as our own IQ data, we were concerned that the children in the BEIP who had experienced institutional care would be at risk for a range of cognitive deficits, particularly those involving memory and executive function. Regarding memory, it is well established that early adversity or stress can elevate the levels of circulating glucocorticoids (for example, cortisol), which in turn can prove detrimental to the integrity of the hippocampus.[13] Damage to the hippocampus has been associated with memory impairments.[14] Accordingly, we predicted that children with a history of institutional rearing would be at risk for memory problems.

When it comes to executive functions, the literature is unambiguous: children with histories of institutional care generally perform more poorly on tests of executive function than do children without such a history.[15] Although we know less about the biological basis for executive functioning than we do about memory (that is, we don't yet know which precise molecular and biological events impair PFC functioning), evidence suggests that we would expect to see deficits in this area as well.

Later in this chapter we discuss a number of executive function tests that can be performed in preschool-age children, but for the most part it is difficult to examine EFs in children under five years or so.[16] Nevertheless, at fifty-four months we performed some simple EF tests, and then at eight years conducted a more sophisticated battery of tests. Specifically, we addressed the following questions:

Do children with a history of early institutional care perform worse on tests of visual memory and on executive functions than do a comparison group of community children? Second, among children with a history of early institutional care, do those assigned to a foster care intervention perform better than their peers who have continued care in the institution? Finally, among those in foster care whose memory and EFs improve, was there a sensitive period that influenced the outcome?

On the basis of previous research by other investigators as well as our own findings at earlier ages (including IQ), we anticipated that we would uncover impairments in planning behavior, in inhibitory control, and in learning and memory among children with a history of early institutional care compared with children without a history of such care. Similarly, given what is known about the protracted course of the development of the PFC, we expected that the FCG would show improvements in EFs compared with the CAUG (although we did not expect the FCG to perform equivalently to the NIG); we were less sure about timing effects, however. As for memory, given the well-known plasticity of the hippocampus and surrounding cortex (which makes learning and memory possible for much of the lifespan), we expected the CAUG to show much worse performance than our other two groups but the FCG to show a dramatic improvement in memory, though again we did not make predictions about timing.

To test these hypotheses, at fifty-four months we performed the "Bear-Dragon" test, which evaluates a form of inhibitory control. At eight years, we performed two ERP tasks, one that directly assessed inhibitory control (called "Go-No-GO"), and a second that assessed response monitoring (Flanker task). We also drew from an automated battery of neuropsychological tasks that Charles Nelson has made extensive use of in the past.[17] We discuss each of these in turn.

Inhibitory Control at Fifty-Four Months: Bear-Dragon

The child's ability to inhibit a pre-potent motor response develops over the preschool period and is seen as an essential skill for adaptive social behavior as well as behavior necessary for learning. A pre-potent response refers to the tendency to repeat a motor act that one has been engaging in earlier in the task; an example might be the need to push the left of a pair of buttons for many trials and then suddenly be required to push the right button. The literature on the effects of institutionalization on children has identified inattention/overactivity as a common occurrence, and we were interested in assessing inhibitory control in children during the preschool period. We used a task developed by Grazyna Kochanska, a developmental psychologist at the University of Iowa, that is an adaptation of the children's game "Simon Says."[18]

In the Bear-Dragon task, a bear and a dragon puppet are used. The child is instructed that whatever the Bear says to do (touch your nose) the child should do, and whatever the Dragon says to do, the child should ignore. The child and the experimenter then play "Bear-Dragon." The child has twenty trials, ten with Bear and ten with Dragon. We counted the number of correct responses across both types of trials. We found that the institutionalized children followed the Bear's directions, indicating that they understood, though they did not inhibit responses to the Dragon's commands.

Foster care children did significantly better, though not nearly as well as same-age typically developing children from the community. Thus, overall, it appears that children assigned to the CAUG show the least inhibitory control, the NIG show the most, and the FCG are somewhere in between. This last point is a theme that resonates throughout our results: that in many cases the foster care intervention boosts performance though not quite to the level of the NIG, at least not through age eight.

Inhibitory Control at Eight Years (Go-No-Go)

Inhibitory control, which is critically involved in learning, as well as in both the cognitive and the social functioning elements of school success (the ability to sit at one's desk and do one's work without distraction; the ability to inhibit saying hurtful things to one's friend), was of particular interest to us, given the high rate of ADHD (attention deficit hyperactivity disorder) that has been observed among children who have been institutionalized (see Chapter 11).

Inhibitory control is the ability to hold back a response in the face of potentially distracting stimuli. Inhibitory control increases throughout middle childhood, and this improvement has been linked to the development of prefrontal cortical neural circuitry.[19]

Among children who spent their early years in institutions, impairments in behavioral regulation have also been noted in a range of executive function measures that tap inhibitory control. Moreover, timing matters: infants who spend less time in institutionalized care perform better on inhibitory control tasks than children who spend more time in deprived conditions.[20] Deficits in inhibitory control among post-institutionalized children have been linked to poorer school achievement in late childhood and worse academic outcomes.[21] Post-institutionalized children within the United States are likewise at a greater risk of lagging behind their non-adopted peers in school and are more likely to need an individualized education plan (IEP) and academic learning services.[22] These data suggest that the early caregiving environment has lasting effects on the development of neural systems involved in inhibitory control.

In our eight-year follow-up we administered a well-known task of inhibitory control, the Go-No-Go task, while recording reaction time, button-press accuracy, and event-related potentials.[23] For this task, we begin by presenting the same stimulus repeatedly, asking the child to push a button on every trial (for example, all "Go" trials). The purpose is to prime the child to respond quickly and on every trial, so that he develops a tendency to push the button (what psychologists refer to as a "pre-potent response"). We then present an interleaved sequence of Go and No-Go trials, in which children are instructed to press a button when they see a certain stimulus and not to press a button when they see another stimulus. In our case 70 percent of the trials were "Go" trials (all letters of the alphabet except the letter "X"), and 30 percent of the trials were No-Go trials (when they saw an "X" they should *not* have pressed the button).

Overall, children who remained in care as usual were less accurate and exhibited longer ERP latencies (slower neural responses) compared with children in the FCG or the NIG. However, children in both the CAUG and the FCG exhibited diminished processing of No-Go cues as inferred from one particular ERP component—the P3—which serves as an index of staying on task. Foster care children also showed enhanced ERP components that reflect error monitoring (the so-called error-related negativity, or ERN, and the error-related positivity, or Pe) compared with children in the CAUG, which suggests that were better at monitoring their performance and more sensitive to errors.

Flanker Task and Go-No-Go

As noted, two executive function skills that may be particularly relevant to risk for psychopathology among post-institutionalized children are inhibitory control and response monitoring. Inhibitory control is the ability to withhold pre-potent actions (that is, those that have built up over time) and suppress irrelevant or distracting information. We measured this ability with the Go-No-Go. Response monitoring (also referred to as error monitoring) is the evaluation of one's own actions after they have occurred. This particular skill of response monitoring works in tandem with other EF skills like inhibitory control by signaling the need to adjust behavior to meet task goals. We measured that skill with both the Go-No-Go and a Flanker task (explained below). The engagement of inhibitory control and response monitoring is guided by the prefrontal cortex and the anterior cingulate cortex (ACC).[24]

The primary component of the event-related potential (ERP) associated with response monitoring is the error-related negativity (ERN).[25] Time-locked to subject responding, the ERN is generated in the anterior cingulate cortex and is thought to be the evaluative signal for action.[26] In a study of internationally adopted children, M. M. Loman and colleagues found that children who had

been in foster care and children who had never experienced any-
thing but care by their biological families (that is, had never been
adopted) had significantly greater neural reactivity on the ERN
compared with children who had previously received institutionalized care.[27]

The Flanker task assesses children's ability to respond to a central
target in the context of distracting stimuli. For this study the target
stimuli consisted of right- or left-facing arrows. Children were instructed to respond as quickly and correctly as possible via button
press to indicate the direction of the middle arrow (right or left).
Congruent trials consisted of stimuli all in the same direction (> >
> > > or < < < < <) whereas incongruent trials had the cen-
tral target in a row of stimuli facing in the opposite direction from
the flanking stimuli (< < > < < or > > < > >). The test trials
consisted of equal numbers of congruent and incongruent trials
presented in a pseudo-random order across two test blocks of 80
trials each for a total of 160 test trials. We recorded EEG during the
task and time locked the subject's button-press response to the
EEG to create the ERN.

What we found was quite interesting.[28] Both children who had
been in the CAUG and in the FCG groups showed an ERN similar to children from the community. However, the magnitude of
the ERN moderated relations between institutionalization and
outcomes only for the foster care group. Children in the foster care
intervention who monitored their performance (that is, had larger
ERN amplitudes) exhibited fewer socioemotional problems compared with foster care children who had smaller ERN amplitudes
(monitored less).

Collectively, these results supported the hypothesis that institutional care leads to decrements in performance in inhibitory control and in error monitoring, and that placement in foster care appears to remediate inhibitory control, at least on this task.

CANTAB

Having demonstrated that institutional rearing had a negative impact on inhibitory control, we were interested to see if other executive function skills might also be affected. Here we drew from the Cambridge Automated Neuropsychological Test and Battery. The CANTAB represents a collection of more than a dozen neuropsychological tests, many drawn from the experimental neuropsychology literature of human patients (for example, those with Alzheimer's Disease) and experimental animals. As a result, unlike a conventional neuropsychological battery used in clinical settings, many of the CANTAB tasks have reasonable neural specificity. That is, research has demonstrated which brain circuits are involved in the performance of these tasks. What makes the CANTAB ideally suited to young children is that the tasks are all done on a touchscreen monitor, in the form of a game, and do not require the child to have very sophisticated motor or language ability.[29]

We focused on the following CANTAB tasks:

1. *Motor screening test.* This test provides a screen for visual, movement, and comprehension difficulties. A flashing cross is displayed on the screen in various locations, and children are instructed to touch it as quickly as possible. Accuracy and response latency measures were recorded.

2. *Delayed matching to sample.* In this classic test of memory, a

child is shown a pattern and then must choose which of four similar patterns exactly matches the original pattern. Multiple trials are presented; on some, the original pattern is obscured before the choices appear, and on others there is a delay between when the sample stimulus appears and when the test stimuli appear. Outcome measures include accuracy and response latency.

3. *Paired associates learning.* This is a classic test of visual memory and new learning. A number of boxes are displayed, some of which contain patterns inside. After a brief delay, the child must identify where each individual pattern was displayed. If the child is incorrect, the trial is repeated. As the child progresses through the task, an increasing number of boxes and patterns are displayed. Our dependent measures include stages completed, number of trials, and number of errors.

4. *Stockings of Cambridge (SOC).* In this spatial planning task, the child must copy a pattern displayed on the screen by moving colored circles one at a time, using the fewest number of moves possible (illustrated in Figure 7.2). Our dependent measures include the number of problems that are solved in the minimum number of moves. Other outcome measures include response latency time and mean moves used.

The top panel of Figure 7.2 illustrates what the final outcome should look like, and the child's task is to replicate this pattern by moving the balls that appear in the lower panel. Only the ball that appears on top of another ball can be moved. How long the child takes to plan each

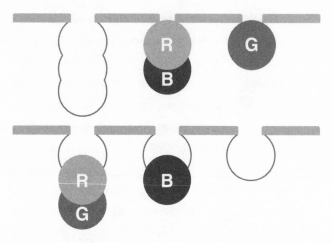

Figure 7.2. Stockings of Cambridge (SOC) spatial planning test. The child's task is to manipulate the balls in the lower panel to reflect the pattern shown in the top panel. Only the ball that appears on top of another ball can be moved. The example shown here is a two-move problem that can be solved by moving the red ball on top of the blue ball and then moving the green ball to the far right location. (R = red; B = blue; G = green)

move, coupled with the number of moves, serve as primary dependent measures. Problems increase in difficulty
up to a 5-move problem.

5. *Spatial working memory.* Here we evaluate a child's ability to
retain spatial information and to manipulate remembered
items in working memory by locating tokens hidden in
boxes. The child is told that after a token has been found in
a box, that box will not contain any tokens in the future.
As trials progress, increasing numbers of boxes and tokens
are presented. Dependent measures include a composite
strategy score, reflecting the child's ability to search
through available items in an organized method, as well as
number of errors for each stage.

For the most part, children enjoy these tasks, though for the more cognitively challenged children, the tests can prove frustrating.

What did we find?[30] First, we observed no significant differences between ever-institutionalized and never-institutionalized children on the motor screening test, which was indeed good news, as it suggests that the basic ability to engage with this battery of tasks was intact (that is, the ability to follow instructions, aim their finger at the screen, touch the appropriate screen location, and so on). Second, and very distressing, we essentially found no difference on any task between the CAUG and the FCG—basically both groups (which we refer to as "ever institutionalized") performed more poorly than the NIG on all other CANTAB tasks (again, except for motor screening), and there were no intervention effects (nor timing effects). This suggests, of course, that institutionalization had a negative impact on these tasks, and that our foster care intervention failed to have a positive effect on the children's performance on these tasks.

Collectively, it appears that ever-institutionalized children experience a decline in performance in both memory and executive functions compared with never-institutionalized children. Moreover, placing such a child in foster care appears not to improve performance in these domains, with the exception of inhibitory control. How might we explain these findings?

First, as discussed previously, the average age at placement in foster care was twenty-two months; thus, it may be that the neural circuits that underlie memory (for example, hippocampus) and executive functions (the prefrontal cortex) are vulnerable to the forms of deprivation associated with institutional care. We find this somewhat surprising, particularly concerning memory circuits, which are thought not to have narrowly defined sensitive periods but instead to be more experience-dependent in nature. However, if the observed effects are due to early perturbation of hippocampal and PFC circuitry, the fact that the intervention failed to improve performance suggests a relative lack of plasticity in these circuits, as well as an early closing of a sensitive period.

Second, it may also be the case that the development of these circuits was perturbed *before* children were placed in an institution. This would explain both the performance deficit observed among the FCG regardless of age at entry into foster care, and the lack of differences between the CAUG and the FCG. However, this suggestion is one we are unable to test, as we have very little prenatal information about our sample. But if it is true, this would support the assertion that children abandoned to institutions are somehow different from those who remain in families.

Overall, consistent with the extant literature, ever-institutionalized children show performance deficits in both memory and executive functions. Moreover, we appear unable to improve such performance by placing children into high-quality foster care, with the exception of inhibitory control, which did improve among the children in foster care.

Language

Language emerges in typically developing infants across the first two years of life. Infants begin expressing themselves with cooing and crying, moving on in the early part of the first half of the first year of life to babbling. By the end of their first year, most infants have spoken their first words, and starting around 18–24 months of age there is an explosion of spoken words by the infant. Emerging during this time is also the basic structure of the child's native language, though learning the rules of one's language takes time and practice.

A child's understanding of language actually appears to emerge even before she can speak. Much of what we know about the conditions or experiences that are important for language development, both receptive and expressive, involve the nature of the input—what the child hears, the frequency of input, the richness of the language stream—the language *environment*. Indeed, research with children of different socioeconomic backgrounds showed that expressive language ability was a function of how much and how often parents spoke to their child and that SES generally has an impact on a variety of neurocognitive functions.[31] Language also plays an important role in social and emotional development. Caregivers' use of language to soothe infants, reassuring them or redirecting their attention, is an important avenue for the emergence of social and emotional regulatory abilities. It is within this context that we assessed the language abilities of children in BEIP.

There are not many standardized assessments of language ability in infants and young children, and those that do exist often rely on caregiver report. We used one, the REEL (Receptive-Expressive Emergent Language) scale to assess language when the children were still living in the institution. Across two primary subscales— expressive and receptive language—we found that the institutionalized children were significantly delayed on all measures of language assessed by caregiver report.[32]

Our project was fortunate in gaining the collaboration of Jennifer Windsor, then chair of the Department of Speech and Hearing Sciences at the University of Minnesota. She has spent considerable time examining the actual language abilities of the children in our study. For example, when the children were thirty months of age, she selected forty children from our sample: ten were from the CAUG group, ten were typically developing community children, and twenty were from the foster care intervention group.[33] Half of these children had been in foster care for at least twelve months and half had just entered into foster care (mean of 2.3 months). Jennifer obtained videos of each child interacting with a caregiver for approximately ten minutes and transcribed the children's utterances. She demonstrated that the children who lived in the institutions and the children who were in foster care for just a very short time showed significantly less lexical grammar production. By contrast, children who had been in our foster care program for at least twelve months looked no different from community controls. And the community children had longer mean length of utterances than even the children who had been in foster care for longer than twelve months. Measures on both the expressive and the receptive scales of the REEL followed a similar pattern.

We followed up a language assessment when the children in the study were forty-two months of age. At that age, children were administered the Reynell Scales of expressive and receptive language. The Reynell contains both expressive language items (for example, object labeling, inflections, clauses) and receptive items (for example, object recognition, following directions). Two Romanian-speaking informants identified which subtests from each scale would be appropriate to administer to the children. In addition, we randomly selected a group of sixty-three children for whom we had a video recording of their interaction with their caregiver. These interactions were transcribed and examined for measures of language complexity. There were two dramatic sets of findings from this study.[34] First, children placed into foster care before fifteen months were indistinguishable from their community same-age peers at forty-two months on all measures of receptive and expressive language. Second, children between fifteen and twenty-four months placed in our foster care intervention experienced dramatic improvements in both receptive and expressive language (although not nearly as great as those placed before fifteen months). There was a strong correlation between age of placement and scores. Also, the linear relation between age of placement and language score was no longer evident after twenty-four months. That is, children placed after twenty-four months of age did not fare well on the language measures.

We again followed up the children at eight years of age. At that time, we administered four expressive language measures including nonword repetition, sentence repetition, written word identification, and average utterance length from a spontaneous language sample. All the tasks and stimuli were developed in conjunction with native Romanian-speaking informants. We found that, even though current placements varied among children in both the CAUG and the FCG groups, children originally placed into our foster care intervention had longer sentences and stronger sentence repetition and written word identification. Children placed in foster care by age two had significant advantages in word identification and nonword repetition: children placed by fifteen months performed the same as their community peers.

These results speak to a number of issues regarding the importance of early experience in the emergence of language. First, they confirm that living in an institution has profound effects on children's language development. Second, they underscore the importance of timing of enhanced environments in facilitating recovery.[35] The human brain most likely expects to receive auditory input that is communicative, and that expectation is probably linked to the

Table 7.2 Intervention and timing effects for cognition, language, and executive function

Construct evaluated	Assessment								
	42 mos.			54 mos.			8 yrs.		
	Intervention effects	Timing effects	Timing observed	Intervention effects	Timing effects	Timing observed	Intervention effects	Timing effects	Timing observed
Cognition	yes	yes	24 mos.	yes	yes	24 mos.	yes	modest	26 mos.
Language	yes	yes	15 mos.		not assessed		yes	yes	24 mos.
Neuropsychological function									
Bear–Dragon		not assessed		yes	no	—		not assessed	
Posner Spatial Cueing		not assessed			not assessed		yes	no	—
Go-No-Go		not assessed	—		not assessed		yes	no	—
Mismatch negativity	no	no			not assessed		yes	not assessed	
Flanker		not assessed			not assessed		yes	no	
CANTAB		not assessed			not assessed		no	no	—

emergence of circuits in the temporal cortex that support receptive language and in the frontal cortex that support expressive language. Finally, the findings support the notion that a foster family intervention provided early enough can ameliorate severe language deficits.

Table 7.2 illustrates the overall pattern of intervention and timing effects we observed in cognitive and language measures. As is evident here, in some domains we observed both intervention and timing effects (for example, language, IQ), in others we observed neither (for example, neuropsychological performance). This unevenness across different domains of development is a theme that will recur across the next four chapters.

Summary

Overall, the cognitive outcomes reveal a very disturbing pattern: children who remain in care as usual suffer from significant declines and delays in all aspects of cognition and language: they have diminished intellectual function as inferred from IQ tests, they show deficits in both executive functions and memory, and they experience profound delays in language. In contrast, children placed in foster care, particularly before they are two years of age, show a dramatic improvement in both IQ and language, though they do not appear to reach the developmental level of never-institutionalized children. Finally, and most distressing, *ever*-institutionalized children display deficits in a variety of executive functions, a pattern that we were mostly unable to reverse by placing children in foster care (regardless of age of placement). We will return to this finding in Chapter 12 when we talk about sensitive periods.

Chapter 8

Early Institutionalization and Brain Development

It is this potential for plasticity of the relatively stereotyped units of the nervous system that endows each of us with our individuality.

Eric R. Kandel, *Principles of Neural Science*

A major focus of our study was the effect of severe psychosocial deprivation on the developing brain and whether transferring a child from a deprived institutional environment to a foster family would change brain structure and functioning. We also hoped to explore whether the timing of the intervention would have an impact on outcome; specifically, is there a period during development when intervention would be most effective? To explore these questions, we used brain-imaging methods suitable for infants and young children—the electroencephalogram and its subset, the event-related potential. Two of us (Nelson and Fox) had been using EEG tools for more than two decades; our combined expertise and the low cost of administering EEG made these tests the obvious choice. By the time the children were eight years old, MRI had become available, and we were able to perform structural imaging on them at this age. Table 8.1 illustrates the brain-based battery we

Table 8.1 Measures used to evaluate brain function and structure

Construct	Measures	Assessment schedule				Source of data
		Baseline	30 mos.	42 mos.	8 yrs.	
Electroencephalogram (EEG)						
	coherence	X	X	X	X	exam of child
	power	X	X	X	X	exam of child
	laterality	X	X	X	X	exam of child
Event-related potentials (ERPs)						
	mother–stranger	X	X	X		exam of child
	facial emotion discrimination	X	X	X		exam of child
	Go–No-Go				X	exam of child
	Flanker				X	exam of child
Magnetic resonance imaging (MRI)						
	brain structure			X	X	exam of child

used, and Table 8.2 shows the timing effects revealed by each of these tools. Before describing our use of these measures among the study group, we first provide a brief explanation of neuroanatomy and the EEG.

White and Gray Matter

The brain contains billions of cells, roughly divided into neurons and glia. The former come in many varieties, with the name of the neuron derived primarily from its shape (for example, pyramidal cell, chandelier cell, and so on). Neurons are the centers of learning, memory, and communication. Glia, which outnumber neurons ten to one, also come in different varieties (some make myelin, a fatty substance that coats the axons of neurons; some help remove debris from the brain). The cell bodies of neurons, along with all forms of glia, are generally referred to as "gray matter," which is easily visible when examining an MRI of the brain. In contrast, depending on location in the brain, the long appendages of some neurons—axons—may or may not be coated with myelin. Myelin helps speed up information transmission (conduction velocity), and on an MRI appears white—hence the name "white matter." Thus white matter refers to parts of the brain that contain myelin (which always coats axons) and gray matter refers to pretty much everything else.

The primary means by which neurons communicate is through the synapse. There are different types of synapses, but perhaps the most common results from the dendrite of one neuron forming a connection with an axon of an adjacent neuron. The brain contains trillions of synapses, each essentially an electrical connection between neurons, made possible by chemicals (neurotransmitters) that are released into the space between axon and dendrite. Information

Table 8.2 Intervention and timing effects for brain function and structure

| | Assessment | | | | | |
| | 42 mos. | | | 8 yrs. | | |
Construct	Intervention effects	Timing effects	Timing observed	Intervention effects	Timing effects	Timing observed
Electroencephalogram (EEG)						
power	no	no	no	yes	yes	24 mos.
coherence	no	no	no	no	not yet analyzed	—
laterality	no	no	no	no	no	—
Event-related potentials (ERPs)						
familiar–unfamiliar face recognition	no	no	—	not assessed		—
facial emotion discrimination	no	no	—	yes (behavioral)	no	—
Go–No-Go	not assessed			yes (behavioral)	no	—
Magnetic resonance imaging (MRI)	not assessed			no	no	—

is transmitted as an electrical impulse along the axon of one cell to the dendrite of an adjacent cell. If the axon is myelinated, it transmits the electrical impulse faster than if it is unmyelinated.

A very common type of neuron in the brain has a pyramid shape and is referred to as a pyramidal cell. The electrical activity generated by synapses of pyramidal cells propagates to the surface of the brain through the space that surrounds neurons—the so-called extracellular space. This activity can be recorded by placing sensors—electrodes—on the scalp surface. Such recordings are referred to as the electroencephalogram, or EEG. A few important points must be emphasized when discussing the EEG. First, the signals that originate in the brain are on the order of millivolts, but by the time these signals propagate to the scalp surface and have gone through brain tissue and extracellular space, these signals are measured at the microvolt level; hence the need to amplify these signals (that is, turn up the volume). Second, the signals that are picked up on the scalp carry with them not only information or signals from neurons but also signals from other physiological processes such as heart rate and thermal regulation; hence these signals need to be filtered so that the actual signal emerges out of the extraneous noise. Third, because these signals travel through extracellular space (rather than along a synaptic pathway itself), it is difficult to infer where they originate when they are recorded at the scalp surface (rather than being recorded in the brain itself). Thus the EEG has relatively poor spatial resolution. Finally, the EEG itself is a complex signal containing many different frequencies. One typically uses sophisticated signal-processing tools to decompose this complex signal into its constituent frequencies—alpha, beta, delta, and so on. It has long been assumed that each frequency reflects a different type of function—for example, alpha activity is associated with consciousness but not active cognitive processing, whereas beta activity is associated with cognitive processing.[1]

EEG

One of the first challenges we faced was building an EEG laboratory. After identifying space in one of the institutions from which we recruited our sample, we needed to purchase the equipment in the United States, bring it to Romania, install it, and then train our staff to use it. Peter Marshall, now a professor of psychology at Temple University, played an instrumental role at this stage in the project, overseeing the construction of the EEG lab and training our Bucharest-based staff. Peter also continuously monitored data collection in the first few years of the project to ensure its integrity.

The EEG records the electrical activity of billions of neurons that populate the brain. To record this activity, we used a Lycra™ stretchable cap with tin electrodes sewn into it. We prepared the sixteen electrode sites by first applying a bit of gel and gently abrading the scalp with the blunt end of a wooden cue tip. Next, we used a small amount of hypoallergenic conductant gel that made contact with the child's scalp (Figure 8.1 shows a study participant wearing the cap). These electrodes pick up the electrical activity that propagates from the brain to the scalp surface. These signals are then amplified (enlarged) and filtered to dispose of non-neuronal electrical activity, such as heart rate, thermal activity, and stray noise from nearby electrical equipment. This procedure has been used with adults since the 1920s and with children since the 1930s. EEG, which is completely noninvasive and painless, provides a measure of the electrical energy generated at certain frequencies in the brain.

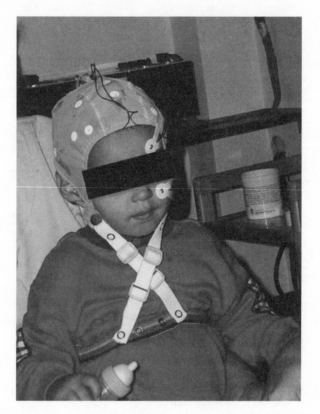

Figure 8.1. The EEG recording system we employed through the first eight years of the project. The yellow Lycra™ cap contains small sensors that make contact with the scalp and, in so doing, permit the researcher to record the EEG. (Photograph by Charles A. Nelson)

We recorded the EEG before the children were randomized and while they all still lived in institutions. After randomization, we again recorded the EEG at thirty months and forty-two months of age, while children were viewing an entertaining visual display: bingo balls rotating in a bingo cage (the goal was to engage the child's attention for a few minutes, standardizing attention to the wheel and minimizing movement).

The complex EEG signal comprises many different oscillatory processes that have specific frequencies and amplitudes, ranging from very slow high-amplitude activity, associated with sleep, to very fast low-amplitude activity, associated with complex information processing. These different oscillations can be broken down into individual frequency bands: alpha activity, which we associate with attention to sensory experience (for example, eyes open, staring at bingo balls), beta activity, which we associate with complex cognitive activity (for example, "Hey, was that the number 7 I just saw?"), and theta activity, which may be stimulus-driven (responsive to memory tasks and some aspects of emotion) but in general is associated with developmental delay as compared with the alpha or beta bands, and delta activity, associated with sleep.

Each of these frequency bands has a distinctive "signature"; for example, we were particularly interested in beta activity (13–20 hz), alpha activity (6–9 hz), and theta activity (3–5 hz). Over the first three years of life, alpha and beta frequencies increasingly contribute to the EEG signal while theta decreases.[2] Changes in both the amplitude and the frequency distribution of the EEG with age are thought to reflect development of the structural integrity of white matter tracts, which again helps speed up the transmission of electrical activity. As white matter increases linearly across child development, so do alpha contributions to the EEG signal. Increased alpha and beta frequency contributions, coupled with decreased theta contributions, have been associated with increased cortical maturity and greater control over attention.[3] One can think of the EEG as a marker of the integrity of the connectivity of brain circuits and one way of tracking cortical maturation.

At baseline, we found large differences in EEG power at particular frequency bands between the children living in institutions and the community controls. The children living in institutions had greater slow-frequency activity (theta) and less high-frequency (alpha and beta) activity relative to the community controls.[4] It was as if someone had turned down the dimmer switch on the higher-frequency electrical activity in the institutionalized children (alpha and beta activity) and increased the less mature (theta) brain activity. This less mature profile was maintained through the initial period of the intervention.

Figure 8.2 illustrates this pattern for theta and alpha activity at baseline, comparing the children in the institutional group with those in the community sample. This figure displays the amount of

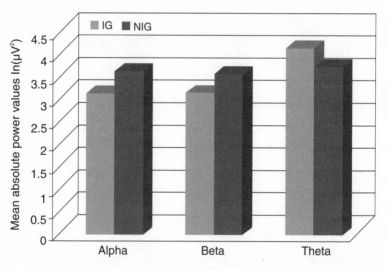

Figure 8.2. Mean absolute power in three frequency bands from institutionalized children (IG; recorded at baseline prior to randomization) and community children (NIG). Note that the institutionalized children show a less mature pattern of brain activity than community children, reflected in less alpha and beta power and more theta power.

brain activity (EEG power) in three frequency bands (alpha, beta, and theta) for the institutionalized children and the community controls. One can see that the institutionalized children had significantly less alpha and beta power but more theta power than the community children.

The institutionalized children show less power in the alpha and beta frequency ranges, whereas the community children show a typical pattern of high power.

Naturally, we recorded the EEG during the follow-up phase of our study as well. However, the intervention effects were at forty-two months relatively subtle. That is, despite no FCG versus CAUG differences, we found a moderate correlation between the amount of electrical activity and the length of time the children in FCG had been placed.[5] By then, children in the FCG had been in their foster care families for around four years. At age eight, there was a strong intervention effect and a timing effect in the EEG data.[6] Children placed into foster care intervention before the age of two displayed a pattern of brain activity that was indistinguishable from that of never-institutionalized children (see Figure 8.3). The children who were taken out of the institution before age two showed a pattern of higher mature alpha activity and lower less mature theta activity. Foster care children also had more significant high-frequency beta activity compared with children randomized to the institution. Indeed, as with the alpha power finding, there was both an intervention and a timing effect for beta power at age eight.

These eight-year data are displayed for each of the sets of electrode sites that were recorded (frontal, central, parietal, occipital, and temporal) and show that the children placed in foster care before twenty-four months of age are indistinguishable from commu-

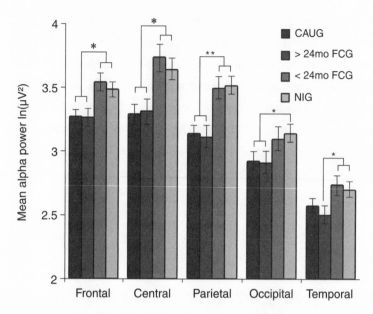

Figure 8.3. Mean alpha power for four groups at eight years of age: children placed into foster care before twenty-four months; children placed in foster care after twenty-four months; children randomized to remain in the institution; and community controls. As can be seen, the children placed into foster care before twenty-four months of age are indistinguishable from community controls.

nity controls at age eight at all these sites, while those placed after twenty-four months are similar in pattern to the children randomized to remain in the institution.

These findings suggested two important conclusions. First, there may be a sensitive period for the development of neural structures underlying increased alpha power in the EEG signal, with amelioration of the environment before two years of age necessary for remediation to occur. Second, though this developmental "catch-up" was made possible by placement into foster care before the age of two, it required years of exposure to foster care to emerge.

We had recorded EEG at thirty and forty-two months and had found little change from baseline to forty-two months. Of course, since the mean age of the children placed into foster care was twenty-two months, the average duration of intervention at thirty months was only eight months and at forty-two months, twenty months. We thought that perhaps this was not long enough to affect brain activity. By eight years of age, children had experienced the intervention and other life events that may have facilitated brain changes, particularly among those who received the initial intervention. Thus, the intervention and timing effects could reflect a subtle combination of age of placement *and* duration of intervention. In other words, the initial period of intervention from a mean of twenty-two months to fifty-four months of age was sufficient to accelerate brain activity in the intervention group. But as time passed and children began to spend more time out of institutional care (as with the child in the care-as-usual group who was removed by the child-protection authorities at age three and reunited with his family), the effects of current life experience began to contribute to development along with the effects of early life experience.

ERP

Whereas the EEG reflects the continuous electrical activity generated by the billions of neurons that make up the brain, the event-related potential (ERP) represents a subset of the EEG. If an individual is continuously presented with a stimulus (for example, picture, sound, touch), over and over again, the brain's electrical activity becomes time-locked to the presentation of the stimulus. Each one of these presentations, coupled with a brief recording period (about 1 second) constitutes a trial. When we average these trials together, emerging from the background EEG is a series of deflections of the signal, referred to as *components*. Each component is thought to reflect a different underlying neural and perceptual/ cognitive operation, with its own unique functional significance. These components take the form of positive and negative "hills and valleys" in the graphic representation of the signal and are labeled using a standard convention: P represents a positive peak and N a negative peak (see Figure 8.4). A number is then affixed to the P or N, and the number is generally taken to reflect the latency, in milliseconds, from stimulus onset to when the peak occurs. For example, the P100 is a positive or upward deflection that occurs roughly 100 milliseconds after the stimulus. The size (amplitude) of a component is thought to reflect the synchronous activity of a large

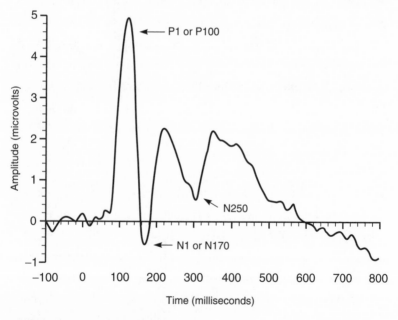

Figure 8.4. Typical adult ERP response recorded over occipital and temporal brain regions.

population of neurons, such that the larger the amplitude, the *more* brain activity. The latency serves as an index of mental chronometry—that is, how rapidly the brain is performing the computations that give rise to a given component. The shorter the latency the more rapid (and efficient) the information processing. Figure 8.4 illustrates a highly stylized ERP.[7]

The morphology of ERP components changes across development, and it has historically been challenging to determine if a component observed in infancy is the same component observed in adulthood. For example, if adults are presented with pictures of faces, they generate a negative (downward) deflection in the EEG signal that occurs about 170 msec after the face is presented. This "N170" is thought to reflect the brain's detection of the structural properties of a face. However, despite the sophistication of face processing early in life, there is no N170; rather, there are three components that appear to have some sensitivity to faces: the P100, the N290, and the P400. The N290 and P400 appear to be manipulated by the same independent variables (such as whether a face is presented upright or inverted) as the N170, and thus are thought to be the developmental precursors to the N170.[8]

This background is important in the context of the ERP studies performed in the BEIP, as we have data from baseline through age eight. In the sections below we describe the various ERP studies we have performed, each designed to examine different underlying questions and/or test different hypotheses.

ERP Studies of Familiar versus Unfamiliar Faces

The characteristics of institutional care make it likely that institutionalized children have unusual experiences with faces compared with children reared in typical families. Such characteristics include a high child-to-caregiver ratio, high caregiver turnover, and limited adult-child interactions within institutions.[9] These factors may lead to more limited exposure to adult faces among institutionalized compared with family-reared children. It is also possible that institutionalized children have access to a more limited range of facial expressions, or have disproportionate experience with particular facial expressions (such as negative or neutral expressions), compared with children reared in typical homes. It is undoubtedly the case that the social interactions during which institutionalized children see faces differ from those experienced by family-reared children; feeding is just one example. For all these reasons, we sought to examine two aspects of BEIP children's responses to faces: their ability to discriminate a familiar from an unfamiliar face and their ability to discriminate different facial expressions of emotions. Both elements of face processing were examined at baseline and at follow-up.

To address these questions, we conducted a simple test: children were presented with alternating images of their primary caregiver's face and the face of a stranger, and ERPs were recorded during this time. Susan Parker, then a graduate student at the University of Minnesota and now a professor of psychology at Randolph-Macon College, oversaw much of this work at the beginning of the project. We examined a series of ERP components across groups, both at baseline and at later ages.[10] We discuss the baseline findings first, followed by the thirty- and forty-two-month findings.

Baseline

We made three major observations at baseline. First, for all ERP components examined, the NIG showed much larger amplitude responses than did the Institutionalized Group (IG) (see Figure 8.5).

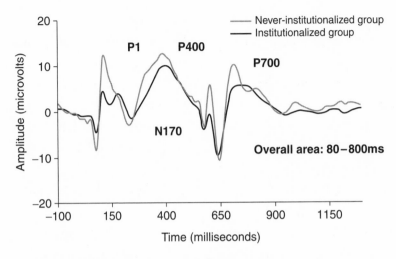

Figure 8.5. Baseline assessment: occipital components.

Second, for the critical component of interest—the negative component, or NC, which reflects allocation of attention—both institutionalized and community children showed the same response, with a larger NC to the stranger's face compared with the caregiver's face. Third, for the Positive Slow Wave (PSW), which reflects the updating of memory, the IG displayed a greater amplitude response to the familiar face (versus the novel face), whereas the NIG showed no ERP difference. Collectively, the amplitude differences we observed between the IG and the NIG are consistent with what we observed in the EEG. In addition, the PSW difference suggests that the children in the IG were attending more to the caregiver's face, implying that this face was not well encoded into memory, whereas the NIG responded equally to both familiar and unfamiliar faces.

Overall, the reduction in amplitudes among the CAUG is consistent with the EEG findings reviewed earlier: institutionalized children generate reduced brain electrical activity. In terms of facial discrimination, it appears that for one component (NC) the CAUG and NIG performed identically, but for another (PSW) the CAUG displayed a very different response from the NIG. This suggests that institutionalization leads to subtle differences in processing familiar and novel faces. The next question was whether such differences carried forward to the forty-two-month follow-up. In other words, if there were slight differences in processing faces early in life, would these be magnified over time so that later in life there would be substantial differences?

Follow-up testing to examine intervention effects when children were forty-two months of age was identical to baseline testing: children were presented with alternating images of their caregiver's face and the face of a stranger while ERPs were recorded. Several findings were noteworthy. First, as we observed at baseline, the CAUG displayed markedly smaller ERP amplitudes. What is important to note about this finding is that at forty-two months of age, only twenty children were still living in an institution, and yet when their data were analyzed per their original group assignment (following our intent-to-treat approach), group amplitudes were still markedly smaller. Thus the hypocortical arousal observed at baseline persisted at forty-two months, suggesting that this brain pattern is a result of early experience (perhaps prenatal as well) and not subsequent life experience. Second, for most components, both the CAUG and the NIG displayed very similar profiles of ERP differentiation of caregiver versus stranger faces; specifically, larger amplitude and shorter latencies to stranger versus caregiver faces. Again, consistent with the baseline findings, this suggests that, much to our surprise, early institutionalization has relatively little impact on discriminating caregiver versus stranger faces. That is, children raised in institutions are as capable of distinguishing familiar and unfamiliar faces as are children who have never been institutionalized. This may be because the development of face processing requires relatively little input to proceed normally.[11] Perhaps just being around other children and a multitude of caregivers is sufficient exposure for the neural circuitry that underlies recognition of faces to develop in a typical fashion.

Finally, we observed that the amplitude of a key ERP component that has face-sensitivity—the P1—was enhanced among children in the FCG compared with the CAUG, though not quite as significantly as for those in the NIG (see Figure 8.6). This result suggests that the foster care intervention had, by forty-two months, begun to normalize this aspect of face processing. This is consistent with the EEG data, in that placing children in families, following early institutionalization, leads to a reduction in the hypo-

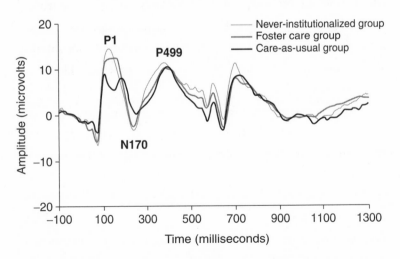

Figure 8.6. Thirty-month assessment: grand averaged ERP waveforms at right occipital electrode (O2).

cortical arousal observed at baseline among the CAUG children. Interestingly, no timing effects were observed, suggesting that normalization occurs regardless of when a child is placed in foster care.

ERP Assessment at Thirty Months of Age

When we review the baseline and follow-up findings, we can draw several conclusions. First, as is the case with EEG, children assigned to remain in institutional care showed a marked reduction in all ERP components. This cortical hypoarousal is of some concern, as it suggests that the brains of institutionalized children (and children who spent the first part of their life in institutional care but eventually went into foster care) develop differently from the brains of children who were never institutionalized. One is tempted to draw a parallel with an earlier observation, which is that the heads of children in the CAUG do not grow as fast nor as large as those in the FCG or the NIG. This pair of findings—reduced head size and reduced brain activity—implies that institutional care leads to a reduction in neurons, a reduction in synapses, and/or disorganization in neural circuitry, any or all of which could be responsible for reduced brain activity.

Second, unlike virtually every other domain observed in BEIP, institutional care appears to have *relatively* little effect on the ability to distinguish familiar from unfamiliar faces. In other words, for only a single component, and only at baseline, did institutionalized children differ from community children—both groups discriminated a familiar face from an unfamiliar face. This, in turn, argues that there is sufficient facial information in the institutional environment to support the typical development of the neural circuitry that underlies face processing. Of course, it remains to be seen if at later ages more complex forms of face processing are observed among the CAUG.

Third, we found that placing children into high-quality foster care (regardless of when such placement occurs) leads to a near-normalization of the brain activity associated with facial-identity processing. This observation, which is consistent with the EEG findings, is encouraging in that it suggests that the harm done to the brain by early institutionalization can be partially remediated by placement in a family, seemingly regardless of the age of placement. Whether complete recovery occurs at later ages remains to be determined.

ERP Studies of Facial Emotion Processing

Experience can have a powerful effect on select aspects of the development of face processing. The development of face processing is most likely an "experience-expectant, activity-dependent process," in which cortical specialization is driven by experiences with faces over the first years of life.[12] Particularly relevant to BEIP are studies that have examined facial emotion processing in children experiencing early institutionalization followed by adoption.[13]

Children with histories of institutional rearing appear to have a number of social difficulties, including deficits in joint attention, indiscriminate behavior, limited social skills, and poor peer relations. These problems may originate in atypical processing of social stimuli such as emotion expression in faces. Accordingly, examining the neural responses to facial expressions of emotion may offer insight into higher-level social deficits in institutionalized children.

At both baseline and follow-up, we adopted a very simple experimental design: infants/children were presented with four facial expressions—happy, sad, angry, and fearful—posed by the same female model. These four expressions appeared equally often, during which we recorded ERPs.

Baseline

Several findings are noteworthy.[14] First, as we observed in the familiar-unfamiliar face task, the amplitude of all ERP components was dramatically reduced among the IG compared with the NIG. This is what we would anticipate since the EEG power amplitudes were similarly reduced. Second, the NIG displayed greater N170 amplitudes to the sad expression, whereas the IG exhibited greater N170 amplitudes to the fearful expression. In addition, the NIG and the IG exhibited greater P250 amplitudes in response to the sad and fearful expressions, respectively. Thus, for the early components, N170 and P250, the NIG and the IG displayed different ERP patterns in response to the different facial expressions of emotion, particularly fearful and sad. Finally, we observed hemispheric differences between the groups, with the NIG showing greater right than left activation among several ERP components and the IG showing virtually no hemispheric differences. The lack of such a laterality effect suggests that the IG children show less cortical specialization for emotion processing than do the NIG. This finding may be consistent with our other observations about reductions in overall cortical activity—thus, it may be that specialization fails to develop in a typical fashion because of a lack of neural substrate to power it.

Follow-Up

This same experimental paradigm was administered at forty-two months following randomization to foster care.[15] As was the case at baseline, at forty-two months, institutionalized children showed markedly smaller amplitudes and longer latencies for the P1, N170, and P400 components compared with the family-reared children (NIG). Moreover, as was the case for the familiar-unfamiliar face task, by forty-two months ERP amplitudes and latencies of children placed in foster care were intermediate between the CAUG and the NIG, suggesting that foster care may be partially effective in ameliorating adverse neural changes caused by institutionalization. The age at which children were placed into foster care was unrelated to their ERP outcomes at forty-two months. Finally, for all three groups of children (CAUG, FCG, and NIG) fearful faces elicited larger amplitude and longer latency responses than did happy faces for the P250 and NC components. This last finding is consistent with past research with typically developing children and suggests that institutional rearing appears to have a limited impact on discriminating facial expressions of emotion.[16]

Collectively, through forty-two months, we observed several differences in emotion processing between institutionalized and never-institutionalized children, though these differences by and large were not great in magnitude. As was the case in the familiar-unfamiliar face task, this suggests that the ability to perceptually discriminate different facial expressions is *relatively* unperturbed by institutional rearing, a finding we have also observed across the first forty-two months of life using a behavioral task of emotion processing.[17] However, as was also the case in the familiar-unfamiliar face task, in the emotion discrimination task we also observed a marked reduction in the amplitude of all ERP components recorded, which was partially remediated by the foster care intervention. This finding suggests reduced brain activity despite intact skills to detect facial expressions of emotion. One way to interpret this result is that the neural network underlying this element of face processing is not as specialized among the IG/CAUG as among the FCG and NIG, owing to more limited exposure to faces.

ERP Studies of Emotion Processing at Age Eight

The two tasks discussed thus far place relatively few demands on higher social-cognitive processes. We were curious what we might find when the children were a bit older if we used a more demanding test of face processing. In other words, how would the children perform on a task that required them to *recognize* facial emotions, not just discriminate one expression from another?

Children were presented with angry, fearful, and neutral faces modeled by eight different female actresses.[18] Each expression was presented equally often. ERPs were recorded and children were instructed to push a button whenever they saw an angry face.

Several interesting findings emerged. First, all three groups were equally accurate at recognizing anger, which suggests that there were no group differences in the ability to follow task instructions. Second, all three groups had a more difficult time recognizing fearful faces, which is consistent with the observation that fear is the most difficult expression to be recognized, not reaching adult levels until adolescence.[19] Third, the NIG and the FCG were both better at inhibiting a button press to neutral and fearful faces than was the CAUG, suggesting either that the CAUG was generally less accurate at recognizing these two emotions *or* that they had more difficulty with inhibiting a response generally (which we found to be true in a Go-No-Go task also used with this sample; see Chapter 7). This finding also suggests that the foster care intervention was effective in improving the ability of previously institutionalized children to recognize neutral and fearful emotions. Finally, we observed no group differences in reaction time.

Regarding the ERP data, several intriguing findings emerged. First, the P1, a component that possesses some face sensitivity, was largest to angry faces for the NIG, smallest among the CAUG, and intermediate for the FCG. In addition, as we observed more generally with ERP amplitudes at forty-two months, this finding also supports the notion that foster care improved the ability to process facial emotion.[20]

In contrast to the P1, the N170 and the P300 were biggest to fear in all groups. So though all three groups were equally good at recognizing anger, the task-relevant stimulus, and showed comparable P300 amplitudes to anger, the fact is that their ERP responses (for both the N170 and the P300) were biggest to fear. This result supports the assertion put forth by Jukka Leppanen and Charles Nelson that this emotion is of high signal value (even as early as seven months of life) and responses to this emotion are unperturbed by institutional rearing.[21]

Although the children in foster care showed improvements in their ability to recognize fearful and neutral faces (to a level of performance comparable to that of the never-institutionalized children), and their P1 to anger was midway between the NIG and the CAUG, we observed no timing effects. On the whole, this finding is identical to what we observed at forty-two months, both for face processing generally and for processing facial emotion.

These findings stand in stark relief to the findings about perceptual discrimination observed at earlier ages. Whereas at earlier ages we found relative sparing in the ability to discriminate facial expressions (as well as differentiate familiar from unfamiliar faces), here we did observe differences among CAUG children in processing facial emotion, both behaviorally and electrophysiologically. Moreover, we observed a number of subtle improvements in processing facial emotion among the foster care group, though their performance did not become comparable to that of the never-institutionalized group. The fact that no timing effects were observed suggests that children placed in foster care improve, though they do so regardless of the age at which they were placed in foster care. Of course, the fact that the average age of placement was twenty-two months urges caution in this interpretation, for it is possible that earlier placement might have revealed timing effects.

The facial processing studies conducted to date show that basic perceptual abilities are only modestly affected by institutionalization, whereas higher-level, recognition abilities are more influenced by early institutionalization. In addition, and in keeping with the EEG findings, it appears that institutionalization leads to dramatic reductions in ERP amplitudes, a deficiency which, by eight years, appears to be remediated among children placed in foster care, regardless of age of placement. Thus, we have an intervention effect in this domain but no timing effect.

Understanding what our EEG findings mean has been perplexing since our initial observations at baseline. Over the years, we have entertained a number of possible interpretations of these data. For example, if smaller head size correlates with smaller brain size, what exactly is it about the brain of the institutionalized child that is smaller? Is it the loss of neurons? Or the loss of neuronal elements, such as axons and dendrites? Or perhaps it is the loss of synapses? Might it be that the normal process of apoptosis—programmed cell death—has gone awry, such that there has been an excessive *pruning* of neurons, as would occur from underuse of circuits? Finally, might there be less white matter—the portions of neurons that contain myelin and that lead to short- and long-distance connections among neurons—which might contribute to our EEG and ERP observations? In other words, might the brain be less well interconnected and, as a result, its electrical activity reduced?

When the children were between the ages of eight and ten we were finally in a position to test some of these hypotheses. We did so by performing structural MRI scans to examine brain structure in detail.

In collaboration with Adina Chirita, a neuroradiologist in Bucharest, we performed structural MRIs (using a Siemens 1.5 Tesla MRI scanner) on seventy-eight randomly selected BEIP children when they were approximately nine years old. Of these, twenty came from the NIG, twenty-seven from the FCG, and thirty-one from the CAUG.[22] Margaret Sheridan, an assistant professor of pediatrics at Harvard Medical School and Boston Children's Hospital, oversaw the processing of these data. We set out with several very specific hypotheses. First, based on the head circumference, EEG, and ERP data, we expected to see an overall reduction in total cerebral volume among the CAUG. What we could not specify was whether this would be due to a reduction in gray matter, in white matter, or in both. Second, based on both human and animal research, we anticipated that the CAUG would show a reduction in the size of the hippocampus. This hypothesis derived from the literature linking chronic stress exposure to elevations in glucocorticoids, which in turn can be neurotoxic to the hippocampus, making it smaller (see Chapter 1). Third, we thought we might observe changes in the amygdala, though the literature is ambiguous as to the direction in which this change might occur—some have reported an enlarged amygdala and others no change.[23] Fourth, we expected to see some changes, most likely in reduced volume, in the corpus callosum (the large bundle of myelinated axons that connects the two hemispheres), possibly because the connections between the hemispheres are sensitive to experiential input. Finally, based on our EEG and ERP findings after placement in foster care, we thought we might see an increase in white matter among the FCG relative to no change among the CAUG or the NIG. This speculation is based on extensive data from rodents, which have consistently shown that increases in myelin can occur when rats are raised in complex (so-called enriched) environments.[24]

Briefly, we observed a reduction in total cortical volume among ever-institutionalized children (CAUG and FCG) compared with never-institutionalized children. We also observed a reduction in gray matter among both the CAUG and the FCG compared with the NIG—in other words, there was no intervention effect on gray matter. In terms of white matter, we observed a reduction in the CAUG compared with the FCG and the NIG, which did not differ from each other. Despite some regional variation, we saw a reduction in the volume of the corpus callosum in the CAUG compared with the other two groups (which do not differ from each other). These last two findings collectively suggest that the foster care intervention is having a beneficial effect on white matter volume but not on gray matter, which is consistent with the experience-dependent nature of white matter (which permits the inference that the CAUG may have fewer synapses). Finally, we did not see any differences across groups in either the hippocampus or the amygdala, and only very subtle changes in one or two regions of the basal ganglia.

We also examined these findings to see whether they could explain the pattern of low-voltage EEG that we had seen since the early assessment points in our study. Using a statistical approach called mediation analysis, we found that EEG power could be explained by differences among the children in white matter volume. Those children with low white matter volumes were more likely to have low EEG power, and, conversely, high white matter volume was related to typical EEG power. This is a critical finding for our research study, since (as will be seen in later chapters) EEG power is a strong predictor of maladaptive outcomes in the institutionalized children (including increased incidence of ADHD symptoms).

Overall, these findings about brain structure may also help to explain our head circumference findings. The CAUG children may have smaller heads and reductions in brain activity because they have fewer neurons—less gray matter—and less white matter in the brain. Moreover, the fact that white matter improved among children placed in foster care—indeed, normalized—is also consistent with our EEG and, in part, our ERP findings.

Conclusion

This chapter has been concerned with the neural sequelae of currently, previously, and never-institutionalized children. The findings suggest that across multiple neurobiological domains, institutional rearing adversely affects brain development. Children in the CAUG have smaller brains and show dramatic reductions in the brain's electrical activity compared with never-institutionalized children. Children assigned to foster care, by contrast, show dramatic increases in their brain's electrical activity and in the amount of white matter it contains, relative to children in the CAUG. However, children in foster care do not show the same remediation in gray matter.

It is important to note that in domains where intervention effects were noted, some, such as EEG, had sensitive periods associated with them, whereas others, such as ERP and MRI, did not. This suggests either that some domains are more experience-dependent than others; or that the *relatively* late age of placement did not permit us to detect sensitive periods that might have been evident at earlier ages. This theme carries forward to other chapters as well.

Chapter 9

Growth, Motor, and Cellular Findings

The failure of infants to thrive in institutions is due to emotional deprivation.

Harry Bakwin (1949)

In Chapter 6 we reviewed studies showing that early psychosocial deprivation associated with institutionalization can lead to a variety of physical, behavioral, and psychological problems. The deficits and developmental delays that result from institutional rearing undoubtedly have their origins in compromised biological and neural development. Whereas in Chapter 8 we focused on the impact of early experience in the BEIP sample on brain development, here we focus on the impact on biology—specifically, physical growth, the prevalence of stereotypies, motor development, and an epigenetic marker of biological health (see Table 9.1 for a summary of the measures we used).

Physical Growth

It is well established that institutional rearing can result in dramatic reductions in a child's height, weight, and head circumference (a

Table 9.1 Measures used to evaluate growth, stereotypies, motor development, and genetics

Construct	Measures	Baseline	30 mos.	42 mos.	54 mos.	8 yrs.	Source of data
				Assessment schedule			
Physical growth							
	height	X	X	X	X	X	exam of child
	weight	X	X	X	X	X	exam of child
	head circumference	X	X	X	X	X	exam of child
Stereotypies							
	disturbances of attachment	X	X	X	X	X	caregiver report/video
Motor development							
	Bruininks–Oseretsky Test of Motor Proficiency					X	exam of child
Genetics							
	telomere length					X	saliva sample from child

coarse metric of brain size). Indeed, by some accounts children lose approximately one month of linear growth (that is, length) for every three to five months they spend in an institution.[1] Ultimately this can lead children who have spent many years in an institution to appear far shorter than their chronological age.

Growth stunting in the context of neglect has two different etiological pathways. In the first, a reduction in caloric intake for the infant, often as a result of inadequate caregiving, leads to a condition known as *failure to thrive.* Often, mothers of infants failing to thrive are depressed, have limited social supports, and live in materially impoverished environments. In the second, less common form of growth stunting, the child's caloric intake is adequate, but the brain fails to produce the hormone needed to stimulate growth. This syndrome, known as *psychosocial dwarfism,* has been shown to be reversible; that is, once these severely neglected children are placed in adequate caregiving environments, their growth hormone levels rise and their growth increases.[2] The stunted growth that occurs in institutionalized children is believed to be more like psychosocial dwarfism, though insufficient research has been conducted on caloric intake or pituitary metabolism in institutionalized children to determine if inadequate caloric intake also plays a role.[3]

In contrast to most previous studies, in the BEIP we were able to observe children's growth while the children were living in an institution, as well as after they were placed in a family. We also were able to examine the relations between growth and quality of caregiving. Finally, we explored the relations between growth and cognition, on the assumption that the more catch-up growth children experience, the greater likelihood that they will also experience gains in cognitive abilities.

We assessed cognitive functioning (Developmental Quotient, Intelligence Quotient; see Chapter 7) and caregiving quality at baseline, thirty months, and forty-two months. Height, weight, and head circumference measurements were obtained on all children at baseline, thirty, forty-two, and fifty-four months, and monthly measurements were also obtained on all CAUG and FCG children from baseline through forty-two months, which provides a level of detail nearly unprecedented in studies such as this.[4]

Overall, the findings were straightforward. First, the institutionalized children were noticeably smaller (for height, weight, and head circumference) at baseline than the never-institutionalized community-control children. Second, institutionalized children who were born small (that is, had low birth weight) were more growth stunted than institutionalized children who were born at normal birth weight (suggesting that being born small is itself a risk factor, one that is magnified by being brought up in an institution). Third, after placement in foster care, children in the FCG made significant gains in height and weight compared with children in the institutional group; however, the rate of growth in head circumference did not differ between the two groups. When we examined duration of intervention, we found that children placed in foster care showed rapid increases in height and weight but not in head circumference during the first twelve months. After twelve months of intervention, 100 percent of the children we placed in foster care were in the normal range for height and 90 percent were within the normal range for weight. No significant change in any parameter occurred between twelve and eighteen months post-placement—in other words, the growth recovery that occurred among children in foster care did so early after the first year of placement in foster care, implying that growth recovery was less an issue of timing of placement than it was duration of intervention (or perhaps an interaction of the two).

We next explored what mediated the gains made among children placed in foster care. We observed that the higher the quality of the caregiving environment (particularly caregiver sensitivity and positive regard for the child) after placement, the greater the catch-up growth in height and weight (but again, not in head circumference). Finally, we observed that the greater the gain in height, the greater the increase in verbal IQ at fifty-four months.

Overall, our findings about children currently living in institutions support the work of others in suggesting that institutional rearing has a very detrimental effect on growth—specifically, on height, weight, and head circumference. Children in the CAUG who were born at low birth weight were particularly affected. In terms of the efficacy of the intervention, we observed dramatic gains in height and weight among children placed in foster care in the first twelve months, though no differences were observed in head circumference between the CAUG and the FCG. This suggests that the intervention had differential effects on these two dimensions of growth—placement in families facilitated height and weight but appears not to have affected head circumference. In addition, we observed that improvement in the quality of the caregiving environment among children placed in foster care correlated positively with improvements in height and weight. Finally, the greater the catch-up in height, the greater the increase in verbal IQ.

Finding that the caregiving environment was associated with catch-up growth was welcome but not surprising news. After all, it has been known for some time that children who are removed from neglectful homes and placed in good families experience rapid gains in growth. However, we were surprised that remediation was limited to height and weight and did not include head circumference. As we saw in Chapter 8, institutionalization has profound effects on the brain's electrical activity, as well as on the amount of gray and white matter (a proxy for number of brain cells and degree of myelinated axons). Specifically, we observed reductions in EEG activity and in both the gray and the white matter; we also found that white matter volume mediated the relationship between early experience and EEG power.[5] Importantly, we found that EEG power remediated by age eight for those children placed into BEIP's foster homes before twenty-four months. Thus the relations between power and head circumference are not direct.

Stereotypies

One of the most striking images to an observer visiting an institution in Romania is the sight of many children rocking back and forth while sitting or on all fours, turning their heads from side to side, or repeatedly bringing their hand to their face, often slapping themselves. Stereotypies are defined as repetitive, invariant movements with no obvious goal or function. They seem to occur when individuals receive diminished or atypical sensory input. For example, at the zoo we see tigers pacing in a cage or elephants swaying back and forth. Such behavior is thought to result from the absence of the kind of sensory input that they would likely get in their natural habitat. The story with young children is similar. When they are raised in species-atypical environments with limited stimulation, they may develop stereotypic movements.[6] It is important to note, however, that other conditions limiting sensory input such as visual impairments or autism are also often accompanied by stereotypies. In these cases, the deprivation is a result of lack of visual stimuli or an inability to accommodate social stimulation—a condition known as disorder-induced deprivation. Typically developing children, particularly when they are young and particularly when they are excited, may occasionally exhibit stereotypies such as hand flapping, but these behaviors are transient and fairly readily distinguished from the more pervasive and intense stereotypies associated with deprivation.

The functional significance of stereotypies among children experiencing deprivation (either of sensory information alone, or more broadly, such as that affecting children raised in institutions) remains unknown, though common interpretations include attempts by the child to provide self-stimulation, coping mechanisms for self-soothing, or expressions of frustration or anxiety.[7] Because the presence of stereotypies implies a maladaptive response by the developing brain, most likely residing in cortical-basal ganglia circuitry, we sought to examine whether the children being reared in institutions would show a higher prevalence of stereotypies than would never-institutionalized children.

There are different ways to assess the presence of stereotypies. One method we used was to ask parents or institutional caregivers to rate how often the child exhibits stereotypies. The advantage to this approach is that caregivers observe the children during typical activities throughout their waking hours and so can capture behaviors that are relatively infrequent in most children. At the same time, though, such ratings involve some degree of subjectivity. Another method we used was to videotape children during their routines and count the number of stereotypies we observed. This approach is more objective than caregiver/parent reports, but it runs the risk of having insufficient samples of time in which to observe intermittently occurring behaviors. The free-running videotapes that we made of the children's behaviors were only about ninety minutes long. Therefore, we decided to use both methods.

We interviewed caregivers about institutionalized children, and parents of the children in foster care or in the community about their children. After inquiring about the presence or absence of stereotypies, the interviewer asked many detailed follow-up questions to learn as much as possible about how often and under what circumstances the child exhibited these behaviors. Raters, unaware of which group the child was from, rated the caregiver responses as rarely or never present, somewhat or sometimes present, or many or often present. We interviewed parents and caregivers of all children at baseline, thirty months, forty-two months, and fifty-four months. We also observed stereotypies during routine activities at baseline and thirty months.

Not surprisingly, the ever-institutionalized children displayed a higher prevalence of stereotypies at baseline compared with the NIG, according to caregiver reports. Nearly 60 percent of the institutionalized children at baseline showed evidence of stereotypies, whereas 21 percent of children living with their parents in the community had stereotypies. Of the 113 institutionalized children who were studied at baseline (we did not include children less than a year of age), more than one-third exhibited many or frequent stereotypies, whereas only one child out of sixty-one in the community group exhibited many or frequent stereotypies.[8] Furthermore, objective ratings during routine activities also demonstrated more stereotypies among institutionalized compared with never-institutionalized community children.

The next question, of course, was whether placement in foster care would reduce the prevalence of stereotypies among formerly institutionalized children. The answer was yes, and at each age of assessment: thirty, forty-two, and fifty-four months, by caregiver report. Children in foster care, who at the time had been placed for an average of eight months, showed significantly fewer stereotypies compared with children who remained in institutions.[9]

Our next question was whether there were timing effects, such that the earlier a child was placed in foster care, the greater the diminution in stereotypies. In fact, the reduction of stereotypies varied as a function of timing of placement, with children placed before twelve months having fewer stereotypies than those placed between twelve and twenty-four months, and children placed after twenty-four months having the most stereotypies.

An association has long been noted between stereotypies and compromised cognitive function, so we were curious to see how this behavior pattern related to IQ and language. We found that for children in foster care (but not for those in care as usual), the greater the prevalence of stereotypies, the lower the child's IQ and language functioning. We are uncertain why this association exists, though we are tempted to speculate that the persistence of stereotypies may serve as an index or marker of an underlying neurological problem, which might, in turn, impact cognitive and language function.

In addition to language and cognition, we also wondered if there was an association between stereotypies and anxiety, given speculation that stereotypies may be a way of managing anxiety. To that end, we used a psychiatric interview of parents and caregivers (Preschool Age Psychiatric Assessment—see Chapter 11). We found no association between anxiety and stereotypies in the children in care as usual or foster care. This is important, because both stereotypies and anxiety were reduced when children were placed in foster care. Our results indicate that foster care most likely exerts its effects on anxiety and stereotypies through different mechanisms.

The fact that more than 60 percent of institutionalized children at baseline (where the mean age was only twenty-two months) showed evidence of stereotypies suggests yet another very deleterious outcome of being raised in an institution. Fortunately, we were able to reduce greatly the prevalence of this behavior pattern if we placed children in families. However, this reduction was tied to age of placement, such that the earlier the placement, the greater the reduction. Nevertheless, though foster care placement led to a reduction in stereotypies, those children who continued to exhibit such behaviors were also more likely to be delayed in language and intellectual abilities.

Thus far, we have shown that children who remained in institutional care were shorter, lighter, had smaller heads, and were more likely to exhibit stereotyped movements than children who began life in an institution but were then moved into foster care (and of course, than those who never spent time in an institution). As we discussed in Chapter 8, these observations foreshadowed our findings from EEG, ERP, and MRI, where we report that the CAUG children showed "under-powered" brains (inferred from reductions in EEG power and ERP amplitude) and a reduction in overall cortical volume (consistent with the notion of having fewer neurons).

Motor Development

A number of reports have suggested that children who have spent time in institutions show delays or impairments in motor development and motor function.[10] Given the dramatic effects we observed on physical development, we were interested to examine motor development in our sample as well. To that end, at the eight-year follow-up, we addressed two fundamental questions: First, is motor functioning compromised in children exposed to early psychosocial deprivation? Second, does a foster care intervention help to improve these outcomes?

We addressed these questions using the Bruininks-Osteretsky Test of Motor Proficiency, a standardized metric of overall motor proficiency.[11] This test focuses on eight domains: fine-motor performance, fine-motor integration, manual dexterity, bilateral coordination, balance, running speed and agility, upper-limb coordination, and strength.

The findings comparing institutionalized and community children were very straightforward—across the board, in every domain assessed, eight-year-old children who had ever been institutionalized performed significantly worse than never-institutionalized children.[12]

It is not clear why institutional rearing would have such dramatic, across-the-board effects on motor performance. One possibility is disuse—children are confined to cribs very early in life, and later, when they are ambulatory, have limited opportunities to practice their motor skills (particularly fine-motor skills). Another is that, as we have discussed, perhaps children who wind up being abandoned to institutions have experienced subtle (or notso-subtle) brain injuries, which might be consistent with the stereotypy findings.

Unfortunately, the intervention was not effective with regard to the motor functions we assessed. That is, at age eight, there were no differences in the scores of children in foster care and those in the care-as-usual group, though both groups lagged significantly behind the community children.

Do these results point to a sensitive period for motor development, meaning that, having been placed "too late" (average age twenty-two months), these children could not recover? Or might it be that the intervention failed to improve performance because foster care parents were focusing most of their efforts on cognitive function and attachment, at the expense, perhaps, of motor skills? It is also possible that more specialized and intensive interventions, such as occupational therapy, might have enhanced recovery. Of course, recovery might just take time, so that catch-up occurs when the children are older.

The fact remains that previously institutionalized children showed dramatic lags in motor skill performance and development compared with never-institutionalized children. As we consider in Chapter 12, this may prove important for other aspects of development later in life.

Telomeres—Insight into Cellular Health

Exposure to adversity and stress has consistently been associated with a range of negative mental and physical health outcomes.[13] Although we and others have clearly demonstrated the psychological and neurological toll of early institutionalization, we were curious as to whether this form of early adversity would also manifest in changes at the cellular level.

To determine if this was in fact the case, we examined telomeres, which are specialized nucleoprotein complexes located at the end of chromosomes. The telomere region of the chromosome appears to promote chromosomal stability—for example, it serves to protect the chromosome through the countless cell divisions that occur throughout the lifespan. Increased oxidative stress or DNA damage due to environmental exposures has been shown to shorten the telomeres. Once reductions in telomere length reach a critical point, cellular senescence is triggered, cell division ceases, and the cell eventually dies.[14]

Telomere shortening is essentially an *epigenetic* process. By this we mean that experience has not changed the structure of the DNA itself, but rather, it has changed how this DNA gets expressed (generally by adding a methyl group or through histone modification). Accelerated telomere shortening has been associated with normal aging and linked to the same negative health outcomes associated with early adversity, including cardiovascular disease, diabetes, and cognitive decline.[15] Recent studies have shown that telomere shortening is also associated with psychological stress, including a history of early maltreatment, mood disorders, self-reported psychological stress, and stress exposure related to being the caretaker of a chronically ill individual.[16] These studies suggest that the acceleration of cellular aging that occurs with psychological stress may represent one mechanism by which early adversity is translated into increased morbidity and mortality across many different health indexes.

We examined telomere length in the BEIP sample when children were between six and ten years of age.[17] We did so by swabbing the cheek (buccal swab) and then extracting DNA from the cells on the swab. We examined the association between average relative telomere length and exposure to institutional care, which we quantified as the percentage of time at baseline and at fifty-four months of age that each child had lived in an institution. We observed that children with greater exposure to institutional care had significantly shorter telomeres in middle childhood. Gender modified this main effect, in which the percentage of time in institutional care at *baseline* significantly predicted telomere length in females, whereas the percentage of institutional care at *fifty-four months* was strongly predictive of telomere length in males.

These are the first epigenetic findings to demonstrate the biological hazards of institutional care. Although a great deal more basic science research is required to interpret telomere shortening across childhood, our results revealed that the longer children spend in an institution, the shorter their telomeres, though the precise relations varied depending on whether the child was a boy or a girl. That is, at baseline, when children were an average of twenty-two months of age, for girls the percentage of their lives that they had been institutionalized was related to telomere shorting in middle childhood, but this was not true for boys. Instead, at fifty-four months of age, the percentage of their lives that boys had spent in an institution was related to telomere length, but this was not true for girls. What does this mean? To be clear, this was an initial study in an area that needs much more exploration, but these results do suggest an important difference between the sexes. That is, our results are compatible with the conclusion that very early stress matters more for girls whereas cumulative stress matters more for boys.

These results are preliminary evidence that institutional rearing has consequences—possibly important ones—at the cellular level. If telomeres can truly be considered a biomarker for cellular health, we would predict that the longer a child spends in an institution, the more compromised his or her short- and long-term health. This is an important area for subsequent research.

Table 9.2 Intervention and timing effects for growth, stereotypies, motor development, and genetics

	Assessment								
	42 months			54 months			8 years		
Construct evaluated	Intervention effects	Timing effects	Timing observed	Intervention effects	Timing effects	Timing observed	Intervention effects	Timing effects	Timing observed
Physical growth									
Height	yes	yes	12 mos.		not assessed			not yet analyzed	
Weight	yes	yes	12 mos.		not assessed			not yet analyzed	
Head circumference	no	yes	12 mos.		not assessed			not yet analyzed	
Stereotypies	yes	yes	12 mos.	yes	yes	12 mos.		not yet analyzed	
Motor development		not assessed			not assessed		no	no	—
Genetics		not assessed			not assessed		yes	no	—

Conclusion

One of the most innovative contributions BEIP has made to our understanding of the effects of early adversity on development has been in the biological domain. Thus, whereas the vast majority of the historical and contemporary literature examining the effects of institutionalization on development has been descriptive, our work has attempted to elucidate the neural and biological mechanisms that help explain the behavioral findings (see Table 9.2 for a summary of our findings to date). In addition, some of our biological findings are interrelated in a way that helps one finding explain another. For example, early on we observed that the CAUG had smaller head circumferences than the NIG. We then observed that the CAUG children also showed diminished EEG activity. Finally, when we performed MRI imaging on the children, we found that the CAUG had smaller volumes of gray and white matter, which likely explains both the head circumference and the EEG data. Similarly, many studies now support the link between early life adversity and later health outcomes; for example, children who are born small for gestational age and grow rapidly during the first year of life are at greater risk for developing heart disease and diabetes much later in life.[18] Our telomere work suggests that growing up in an institution can compromise a child's cellular growth. We hope in the future to examine the functional consequences of this finding—for example, might the CAUG be more likely than the FCG and the NIG to get sick or eventually suffer from adult-onset diseases such as cardiovascular disease, metabolic syndrome, and so on? We will have to wait for the answers to these questions.

Chapter 10

Socioemotional Development

Infants deprived of their mothers in the first year of life for more than five months deteriorate progressively. They become lethargic, their motility retarded, their weight and growth arrested. Their face becomes vacuous; their activity is restricted to atypical, bizarre finger movements.

Rene Spitz, "Hospitalism" (1945)

A striking feature of the institutions in which we worked was how remarkably subdued and somber the children appeared. Fifteen toddlers between the ages of twelve and thirty-six months would be scattered about a large and mostly empty playroom. Usually one caregiver was with them, sometimes interacting half-heartedly with a particular child, but often sitting and paying little attention to the children.

There were differences among the children to be sure. A few unhesitatingly approached us with open arms, vocalizing and smiling. Others were arrayed around the periphery of the room, engaging in rocking motions while sitting or lying on their backs. Many wandered aimlessly with glazed expressions that were unnerving for their lack of emotional expressiveness. Some of the children became distressed, perhaps in response to being hit by a peer, or sometimes for no apparent reason.

The children rarely sought comfort from a caregiver when they were distressed. Mostly, they just cried. On the few occasions that caregivers offered comfort, either because we were there or because the child's crying was hard to ignore, the child was not readily soothed. The children seemed to have no expectation of being comforted nor any familiarity with the distress-comfort-relief-resumption-of-play cycle that occurs again and again in typical child-caregiver interactions. Moreover, the chief form of social interaction among these children appeared to be brief bouts of aggression—sometimes disputing possession of one of the few available toys or sometimes seemingly unprovoked. There was little sustained play among the children.

Abnormalities in social and emotional behavior in children raised in institutions have been described for more than a hundred years.[1] They were the most striking atypical behaviors in the descriptive studies of Rene Spitz and William Goldfarb in the mid-twentieth century (see Chapter 6). Gripping visual images of young children crying without comfort and later staring with vacant passivity and emotional detachment were notable in Spitz's movies of a foundling home in Central America and in James and Joyce Robertson's movies of young children placed in residential nurseries in London in the 1950s.[2]

These descriptive studies were important in drawing attention to the significant effects of psychosocial deprivation on socioemotional development, but they did not systematically investigate social and emotional behavior in institutionalized children. Several, more contemporary studies have demonstrated profound social and emotional abnormalities among children adopted out of institutions.[3] Our study extends these findings by examining children both while they were living in institutions and following intervention with placement into foster care. Our goal was to determine experimentally how early experiences of deprivation and subsequent placement in foster families affected the children's social and emotional behavior.

One of our first goals in BEIP was to use standardized and validated assessments of social and emotional behavior, including attachment behaviors, to study young children being reared in institutions. In doing so, we were cognizant that we were using measures in settings that were very different from the settings in which they were developed. Therefore, we had to remain aware of the assumptions underlying these measures. For example, the traditionally accepted measure for assessing attachment in young children is a procedure known as the Strange Situation. It was designed to assess qualitative patterns of attachment between young children and their parents. And yet, we used the measure in a setting in which many of the children might not have even formed attachments to their caregivers. In addition, we relied on caregiver reports of children's behaviors, as we often do in studies of young children, realizing that, compared with parents (biological or foster), these institutional caregivers were less likely to know the children and less likely to be psychologically invested in them. Nevertheless, the extreme population represented by institutionalized children allowed us to gauge both the profundity of the effects and the plasticity of the attachment system to rebound in the face of adversity.

On the basis of our own observations as well as previous research, we assumed that because young children raised in institutions have reduced exposure to appropriate social stimulation and contingent sensitive and responsive caregiving, their social development would be compromised. We were particularly interested in three areas likely to be affected: the recognition and expression of positive emotions; attachment behaviors; and peer relationships and social skills.

Positive Emotional Expression

Especially in younger children, interpretation of how funny or threatening or interesting unfamiliar stimuli are depends on the ability to understand the situation and to identify emotions in the facial expressions of others. In the absence, or at least the scarcity, of experiences of these kinds of situations, children may not develop the capacity to experience and express positive emotions.

Measurement of Emotional Expression

Rather than merely observe children's emotional expressiveness during day-to-day interactions, we elected to administer a procedure designed to *elicit* positive emotional expressions from children. There were two reasons for this decision. First, the number of hours of observation we needed to see enough positive emotional expression to be able to detect individual differences was unknown but likely to be many. Second, and more important, we were interested in assessing the children's *capacity* to express positive emotions, since it might well be that the drab environments of institutions affected expression rather than capacity.

To assess capacity for positive emotional expression, we selected two tasks from an assessment battery used to identify individual differences in children's temperament.[4] Emotional expression is considered an important component of temperament, and we used tasks from a measure called the Laboratory Temperament Assessment Battery (LABTAB).[5] This measure was designed to assess individual differences in young children's emotional expressivity. We chose two tasks, puppets and peek-a-boo (both thought to elicit positive emotions). Both tasks were administered at the baseline assessment and again when children were thirty and forty-two months old. The first, slightly modified task we used involved having children sit at a small table on the lap of their institutional caregiver (in the case of institutional children) or biological parent (in the case of community children) or the foster care parent (for those thirty- and forty-two-month-olds randomized to the intervention). Each child faced a research assistant who had a pair of puppets, a fish and a rooster, on either hand. Using different voices for each of the puppets, the research assistant introduced the puppets to the child and carried on a "conversation" with them. The puppets said they were ready to "have some fun" with the child and cautiously tickled the child. In the second task, the research assistant played two rounds of peek-a-boo by covering her face with her hands before smiling at the child and saying, "peek-a-boo." The delay in the second round was a bit longer to create more anticipation. The puppets procedure lasted for two to three minutes and the peek-a-boo task about a minute. The caregiver/parent was instructed not to speak to the child and to remain emotionally neutral during both interactions. This instruction was necessary because young children take their cue from adult caregivers about how to respond emotionally to unfamiliar or ambiguous stimuli. We wanted to be sure the child was responding without a "hint" from the caregiver.

The children were videotaped during these tasks, and their emotional responsiveness was later coded by raters who were unfamiliar with the child's group. Coders recorded instances of specific behaviors, and these were transformed into overall scores of positive emotional expression, negative emotional expression, and attention to task.

Emotional Expressions in Children Living in Institutions

At baseline, the children ranged in age from six to thirty-one months, and the tasks were repeated at thirty months and forty-two months. As always, those children in either the care-as-usual or the foster care group who were older than thirty months at the time of their baseline assessment were brought back for reassessment after one month in their post-baseline placement. This means that at the time of the thirty-month follow-up, the foster care children had had anywhere from one to twenty-four months of placement with a family, and at forty-two months of age, they had had anywhere from twelve to thirty-six months with a family.

Results were so similar for behavior during the puppets and the peek-a-boo tasks that we combined the children's scores in the two episodes. At baseline, we found that the community children, living with their parents, expressed positive emotion more and paid more attention to the task than did the institutionalized children. Institutionalized children displayed more negative expressions of emotion during the two tasks.

Intervention Effects on Emotional Expression

To assess the effect of foster care placement, we looked at the results at thirty and forty-two months in the same procedures and again collapsed them into overall scores reflecting behaviors in both tasks. Before assessing results of the intervention, we needed to determine if—at baseline—the children who would continue to receive care as usual and the children who would be placed in foster care showed any differences in their behavior on these tasks. They did not. At thirty months and at forty-two months, however, the differences were striking. The children in foster care had significantly more positive emotional expressions and better attention to tasks than did the care-as-usual children. The children in foster care not only displayed more positive expressions of emotion at each age, but the change over time for these children from the baseline assessment was significantly more marked than for children receiving care as usual.[6]

In addition, the children in foster care showed at least as much positive emotional expression as did the community children. This means, in addition to demonstrating a large difference between the foster care and the care-as-usual groups, we also showed that the foster care children had recovered from the limited capacity for positive emotional expression before randomization and intervention. This kind of complete recovery was rare in BEIP because on most measures we administered, the foster care children's performance, though superior to the CAUG's, did not match the community children's performance. But at least in this one area, the recovery was early (already evident after an average of eight months in foster care) and complete. Because there were no timing effects, we concluded that the capacity for emotional expression remains open to experience for at least the first two and a half years of life and that it responds very quickly to positive changes in the emotional climate or context.

Attachment

BEIP afforded an opportunity to study the attachment behaviors of young children living in institutions with standardized assessments, and to examine the effects of a family-based intervention on these behaviors. One goal of our study was to determine if validated measures of attachment replicated observations in previous descriptive studies of serious attachment disturbances among many children raised in institutions. Another goal was to examine young children's capacity for recovery, after having been raised at least initially in what we presumed were limited opportunities to form selective attachments. Caregiving quality has long been asserted to be necessary for the healthy development of attachment, so institutional rearing provided a stark test of that position. Finally, we wanted to determine if aberrant attachment was importantly related to the emergence of psychopathology, as it is asserted to be in noninstitutionalized children.

The concept of attachment has a long and complex history in the fields of developmental psychology and psychiatry. Understanding exactly how and why infants grow to select specific individuals as sources of comfort and support has been of interest for many years. Psychoanalytic theory was the first psychological approach to describe the importance of early caregiving infant experiences, linking attachment to oral gratification. This approach was later instantiated in behaviorist descriptions which proposed that the caregiver-infant relationship could be understood within the context of secondary drive theory. Both the psychodynamic and the behaviorist approaches have found little support for the idea that the development of selected preferences by human infants in the first year of life for specific individuals is based solely on an infant's linking source of food to attachment. The famous experiments of Harry Harlow in the 1960s with infant monkeys and the wire mothers who dispensed food and the cloth "mothers" who did not demonstrated that young monkeys preferred "figures" that provided contact comfort (the cloth mother).[7]

John Bowlby, a psychoanalyst, constructed attachment theory by integrating what was known at the time about the development of infant cognitions, perception, and ethological approaches to understanding mechanisms involved in the formation of the attachment bond.[8] The empirical research of Bowlby's student Mary Ainsworth and her students operationalized many of Bowlby's ideas and provided a foundation for the measurement and assessment of this attachment relationship, as well as an understanding of the processes over the first year that lead to individual differences in the quality of the attachment relationship between infant and mother.

Ainsworth and her students studied a small group of mothers and infants in Baltimore, Maryland, and carefully observed the caregiving behaviors that were exhibited, linking these behaviors to the child's behavior when the caregiver was under mild to moderate stress.[9] They demonstrated links between behaviors of infants during the first year of life observed in the home with behaviors at twelve months in the lab, all directed toward their mothers. Caregivers who had been sensitive to their infants' signals of distress or positive affect and who provided responsive contingent care when the infant was distressed were likely to have infants who exhibited proximity-seeking and contact-maintaining behaviors, and they were calmed by the mother's presence. Caregivers who often ignored their infant's signals of distress or who were intrusive were unlikely to have infants who were securely attached; instead, these attachments were deemed to be "insecure." In both types, the infants had formed attachment relationships to their caregivers, but the quality of the relationship varied as a function of the type of caregiving that the infant received over the first year of life.

But what of infants whose experiences are so aberrant as to fail to facilitate even the most basic attachment responses? In institutional settings, most infants do not have a single caregiver nor even consistent multiple caregivers. Infants who grow up in settings deprived of contingent, sensitive, responsive care from early infancy provide an extreme example of the limits to the emergence of attachment, an experience-expectant process.

Attachment is defined in different ways, but generally it refers to a young child's tendency to seek comfort, support, nurturance, and protection from selected adults who are providing care to them. In species-typical circumstances, by seven to nine months of age, infants begin to exhibit clear preferences for certain caregivers when they are frightened or distressed. New behaviors appear as these attachment relationships develop, namely, infants protest separation from selective caregivers, and they exhibit reticence that had not been present before with unfamiliar adults. As infants begin to move around, either crawling or walking by age twelve months, they begin to display behaviors that involve seeking proximity and contact with their preferred caregiver. A constellation of these behaviors toward caregiving adults, including seeking comfort, proximity seeking, and sometimes distress upon separation, suggests that an infant is "attached." Hallmarks of the emergence of attachment relationships are when infants begin to display stranger wariness to unfamiliar adults and separation protest when caregivers leave them for a brief period of time. The process of forming an attachment centers around the caregiver's providing sensitive and responsive care when the infant is distressed or feels fearful. Exactly how many such interactions (quantity vs. quality) are necessary to promote attachment is unknown.

For infants reared in institutions, we were not sure whether the attachments they formed might be atypical and tenuous rather than healthy and robust. Developmental research had established methods for determining qualities of attachment between young children and their caregivers, and we were eager to use these measures to see what they might reveal about this important developmental domain.

Measurement of Attachment in Children Living in Institutions

Because it is considered the most accepted measure of attachment in early childhood, we used the Strange Situation Procedure to assess the institutionalized children.[10] For this assessment, an infant between the ages of one and two years is observed with a parent/caregiver in a series of brief separation and reunion episodes. Attachment theory posits that young children balance an inherent need to venture away from attachment figures to explore with a need to seek physical proximity to an attachment figure in order to feel more secure in times of distress. During the Strange Situation, we would expect a child to look more comfortable and to be more motivated to explore in the presence of an attachment figure. During episodes that involve separation from the attachment figure, we would expect to see the child exhibit overt distress or at least diminished motivation to explore the environment. In coding the Strange Situation, infant behavior, particularly during reunion episodes, is noted for degree of proximity seeking and contact comfort, as well as calming of any distress that the infant might express in the prior separation episode. The organization of the child's responses during these separation and reunion episodes allows trained coders to determine the quality of the child's attachment relationship to the caregiver.

Differences in how infants organize their attachment and exploratory behaviors, especially during reunion episodes, can be reliably classified into different *patterns* of attachment. Infants who become distressed by separation from the caregiver, express their distress readily to their caregiver, seek comfort immediately on reunion, respond readily to soothing, and resume exploration after settling are classified as *securely attached* to the caregiver. Roughly 50–60 percent of typically developing infants exhibit secure attachment to their caregivers. Infants who become distressed by the separation but cannot be soothed by the caregiver on reunion and remain distressed are classified as *ambivalent/resistant,* and these constitute 5–10 percent of attachments in low-risk samples. Another 10–15 percent of infants who display little if any reaction to the separation and actively avoid or ignore their caregiver on reunion are classified as *avoidant.*[11] In infants who have formed attachments, these patterns of attachment are often relationship-specific (that is, they may be different with different caregivers). The idea is that an infant may have multiple attachment relationships, each with its own history of caregiving and soothing in times of distress.

A fourth pattern, *disorganized* attachment, describes young children who lack an organized strategy for obtaining comfort, instead exhibiting aberrant behaviors that are poorly organized or fearful.[12] Roughly 15 percent of low-risk infants but as many as 80 percent of maltreated infants exhibit disorganized attachments with their caregivers.[13] The greater the number and severity of risk factors within parent or child the greater the probability of disorganized attachment.[14] What's important to note here is that in both the different patterns of response and quality of attachment, and even with disorganized attachment, infants are assumed to have formed some type of attachment relationship with a caregiver. It is not that they are "unattached."

Using a slightly modified Strange Situation, researchers also have identified patterns in preschool children analogous to those in infants.[15] Secure, avoidant, and resistant/ambivalent (dependent in preschoolers) are descriptively similar to the same patterns in infancy. Disorganized attachment in infants may be manifest as disorganized or *controlling* attachment in preschoolers. In addition, another aberrant pattern known as *insecure/other* has been identified in high-risk preschoolers.

Secure attachments in infants and preschoolers are thought to be protective factors, that is, they reduce the probability of maladaptive outcomes, especially in high-risk samples of children. Disorganized, controlling, and insecure/other attachment patterns in preschoolers, by contrast, are those most strongly predictive of risk for subsequent psychopathology.[16]

These patterns of attachment in infants, toddlers, and preschoolers have proven valuable for understanding the quality of relationships between parents and young children in high-risk and low-risk samples.[17] Considerable research on qualitative differences in attachment has shown modest links between quality of caregiving and type of attachment, with sensitivity and responsivity of the caregiver associated with secure attachment.[18] Furthermore, in extremes of caregiving adversity, such as maltreatment, exposure to interpersonal violence, or maternal substance abuse, aberrant forms of attachment are more likely.[19] For all these reasons, we wanted to use the Strange Situation Procedure to assess attachment in young children living in institutions.

Of course, the Strange Situation was developed to assess qualitative patterns of attachment in young children who have formed emotional bonds with their caregivers. We could not be certain that children living in institutions actually had formed attachments to their caregivers. Nevertheless, we decided to use the procedure so that we could compare our results with what had been demonstrated in children in diverse circumstances and cultures to see what

the assessment suggested about attachments between institutionalized children and their caregivers.

An immediate question was which caregiver to select to interact
with the child in the Strange Situation. In a setting in which children have limited contact with large numbers of caregivers, it is not
clear who the putative attachment figure might be. We decided to
ask the staff in institutions if the child seemed to have a favorite
caregiver. It turned out that staff consensus was generally quite reliable about favorites. If the child did not have a favorite, we asked
that a caregiver who worked with the child regularly and knew the
child well participate in the procedure.

It was important that during the interactions we videotaped, the
caregivers appeared indistinguishable from mothers. All caregivers
dressed in street clothes, and the videotapes were coded by attachment experts who were unaware of group status. On the basis of
established criteria comparing the balance between their exploratory and proximity seeking, infants were assigned to one of four
major patterns: secure, avoidant, resistant/ambivalent, or disorganized. On a subset of the tapes, two attachment experts compared
their coding data and were found to be similar.

Patterns of Attachment in Children Living in Institutions

BEIP was the first study of patterns of attachment in Romanian
infants and toddlers, including those living in institutions, and the
results we obtained were striking. Among the children living with
their families in the community of Bucharest who had never been
institutionalized, 74 percent had secure attachments, 4 percent had
avoidant attachments, and 22 percent had disorganized attachments
to their mothers. The percentages of secure and disorganized patterns are slightly higher than we might expect within a typically

developing sample in the United States, but it is not unusual for a single study to have distributions with these kinds of minor variations. Among the institutionalized children only 18 percent had secure attachments, 3 percent had avoidant attachments, 65 percent had disorganized attachments, and 13 percent of the children had such limited attachment behavior that they were designated *unclassified*.[20] This last designation is exceedingly uncommon and has never been reported in children living in families.

These remarkable differences between institutionalized children and noninstitutionalized children were similar to what another group of investigators reported after BEIP began from their study of Greek infants raised in institutions.[21] A subsequent study of infants and toddlers raised in Ukrainian institutions revealed similar distributions of patterns of attachment.[22] The similarities of findings in these other two studies regarding attachment contrast with the IQ scores, which were much lower in BEIP than in the Greek or Ukrainian children. This difference suggests that attachment may be a more sensitive indicator of psychological disturbance, whereas IQ may index degree of deprivation. An important older study of residential nurseries in London in the 1970s also described modest effects on IQ but profound effects on young children's attachment.[23] The convergence of our results with other studies increased our confidence that the use of the Strange Situation in the Romanian institutional context had yielded meaningful results.

We were also curious about whether we could predict which variables were important for developing an organized attachment to a caregiver in the extraordinary setting of an institution. Of the many variables we tested, only quality of caregiving was significantly predictive. An increase of one unit in quality of caregiving was associated with a 29 percent increase in the odds of an institutionalized child's having an organized attachment. When children from the unclassified group were included in the disorganized group and the same predictors were used, an increase of one unit in quality of caregiving was associated with a 30 percent increase in the odds of an institutionalized child's having an organized attachment.[24]

Degree of Development of Attachments in Children Living in Institutions

Despite the remarkable differences in patterns of attachment between institutionalized and community children in Romania, these results do not adequately convey the profundity of differences in attachment in these two groups. Indeed, even more extensive differences in attachment between the institutionalized and the community groups became readily apparent.

Elizabeth Carlson, a recognized attachment expert at the University of Minnesota, was the primary coder of the attachment tapes. She knew only that the study involved Romanian children, and she observed only the twenty-one-minute Strange Situation for each child. In the course of coding the BEIP tapes, she noted what she believed to be incompletely formed attachments between the child and the caregiver.

What gave Carlson pause was that many of the children in the tapes she was coding exhibited behaviors that were very unusual; moreover, sometimes children failed to demonstrate any preferences at all. In response, Carlson developed a 5-point rating scale designed to assess the degree to which an attachment had formed between child and caregiver. The scale described a range of child behaviors that did not fit the traditional pattern scheme but appeared to reflect the degree of, or stages in, attachment formation.[25]

In the Carlson scheme, ratings of "5" indicated fully formed attachments, consistent with traditional patterns of attachment, including the category "disorganized." That is, a child could be "classified" as avoidant or resistant and insecure or even disorganized but still attached and would receive a 5. Ratings of "4" indicated evidence of formed attachments accompanied by extreme or pervasive behavioral anomalies (beyond the scope of traditional disorganization coding). Ratings of "3" were used when there were fragmented or incomplete sequences of attachment behavior that were directed more toward the caregiver than toward the stranger. The next level, "2," indicated only a marginally discernible difference in the child's behavior toward the familiar caregiver, with isolated attachment signals and responses. Finally, children who exhibited no discernible attachment behaviors were assigned the lowest rating of "1." Carlson's ratings were compared with those of another attachment expert, Alan Sroufe, now emeritus professor at the University of Minnesota. Their ratings showed high levels of agreement about patterns of attachment as well as the ratings of the degree to which attachment had formed on the 5-point rating scale.

We were careful to include in all analyses of attachment only children who we were sure had attained a cognitive age of eleven months. Because coding of the child's behavior in the Strange Situation depends upon his ability to differentiate between familiar and unfamiliar adults and organize his behavior in response to mild stress, we wanted to be sure that the children we assessed had the cognitive and motor maturity sufficient to assess their attachment behaviors. As a result, we ended up including 95 of the 136 institutionalized children as well as 50 community children.

The results of this analysis were remarkable. With no knowledge about the children whose behavior they coded, the raters identified all 50 of the community children as level 5—having fully developed attachments and recognizable behaviors. In contrast, they identified only 3 of the 95 children in the institutionalized group who displayed behaviors that suggested they had formed attachments. In fact, 62 of the 95 children (65 percent) in the institutionalized group were level 3 or below, meaning that they were displaying anomalous behaviors only minimally related to attachment. Ten percent of the institutionalized children had no discernible attachment behavior whatsoever.[26]

These results were not entirely surprising given the extraordinarily deprived caregiving setting that institutions represent, although this type of analysis had never been conducted previously. A subsequent study of institutionalized children in Ukraine replicated our findings of incompletely formed attachments, though with somewhat more fully formed, or level 5, attachments among institutionalized children than we found (24 percent versus 3 percent).[27]

Another point of interest emerged from comparing ratings of the degree to which attachment had formed and the forced classification of patterns of attachment. First, all 9 children with ratings of "1" (no discernible attachments) were rated as unclassifiable. Second, 7 children (29 percent) whose attachment was rated as level 2, 7 children (24 percent) whose attachment was rated level 3, and 3 children (10 percent) whose attachment was rated level 4 were securely attached. This means that the limited attachment behaviors these children displayed most resembled secure attachment behaviors—things like displaying distress during separation and seeking comfort from the caregiver rather than from the stranger. But clearly, they should not be considered comparable to secure attachments in the community children, all of which were fully formed.

To try to clarify what led children to have either absent or incompletely formed attachments, we examined several possible variables. The only factor that we could identify was quality of caregiving. It appears that what was missing in these children's lives were sensitive, responsive caregivers who could provide them with contingent positive social interaction. In the absence of such persons, these children did not form attachments to their caregivers.

Disorders of Attachment in Children Living in Institutions

As we have seen, under typical rearing conditions, all infants become attached to caregivers. In more extreme rearing conditions, such as institutions for young children, attachment may be seriously compromised or even absent. The work we have described up to now on patterns of attachment derives from the tradition of basic developmental psychology. Another tradition in attachment, also based in John Bowlby's work, has been more rooted in the clinical literature on children raised in institutions. This literature has defined a psychiatric disorder involving aberrant attachment behaviors.

Reactive attachment disorder (RAD) describes a constellation of aberrant attachment behaviors and other social abnormalities that are believed to result from social neglect and deprivation. Two clinical patterns have been described: an emotionally withdrawn/inhibited pattern and an indiscriminately social/disinhibited pattern. In the emotionally withdrawn/inhibited pattern the child exhibits limited or no discernible attachment behavior even when he or she is distressed. In addition, the child fails to seek comfort for distress or to be comforted when distressed and shows emotion regulation difficulties such as unexplained outbursts of irritability, fear, or withdrawal. In the indiscriminately social/disinhibited pattern, the child exhibits lack of expected social reticence with unfamiliar adults, fails to check back with the caregiver in unfamiliar settings, and demonstrates a willingness to "go off" with strangers.[28]

Although systematic study of attachment disorders is quite recent, these disorders have been described in young deprived children for more than half a century. BEIP provided an opportunity to examine a number of questions that had not been examined in any earlier studies. First, it allowed us to evaluate attachment disorders systematically among currently institutionalized children, using contemporary measures. Most of the previous work had been descriptive studies from the mid-twentieth century. Second, it provided an opportunity to evaluate the natural history of attachment disorders within a high-risk sample. Third, it afforded us an unprecedented chance to examine interrelationships between the developmental patterns of attachment—secure, avoidant, resistant, and disorganized—and clinical disorders of attachment. Fourth, it let us examine individual differences in genes, so called polymorphisms, as vulnerability factors for the disorder. Fifth, much as with attachment classifications previously described, it enabled us to determine if foster care—putatively an enhanced form of caregiving—is an effective intervention for disordered attachment. No intentional intervention study for attachment disorders had ever been conducted, much less as a randomized controlled trial. Finally, as with other outcomes in BEIP, we were also able to examine effects of timing of placement in foster care and determine the degree to which timing mattered. These were all compelling reasons for including attachment disorders as a BEIP outcome of interest.

Descriptions of attachment disorders originally appeared in studies of institutionalized and maltreated children. One study published in 1975 was particularly influential.[29] In London in the late

1960s and early 1970s, young abandoned children were placed in residential nurseries. These nurseries were smaller than typical institutions, housing fifteen to twenty-five children on a baby unit or in mixed-aged groups comprising toddlers and preschoolers. The residential nurseries also were better-equipped institutions than many described in earlier reports. They were generally sufficiently staffed, and the children had adequate toys and books and a schedule of activities that were stimulating and age appropriate. In fact, by the age of three, the IQs of these children were in the average range. As in other institutions, however, the staff worked rotating shifts, and importantly, they were explicitly discouraged from forming attachments to the children. The rationale was that the children would be adversely affected if such relationships were ever disrupted.

The primary investigator for this study was Barbara Tizard, a London psychologist interested in adoption and in institutional rearing. She studied sixty-five young children who were abandoned and placed in residential nurseries within the first few months of life until at least their second birthdays. Between the ages of two and four and a half years, twenty-four of the children were adopted and fifteen were returned to their biological parents. This left twenty-six children who had remained in institutions from the earliest months of life through four and a half years.

Of these twenty-six, the nurses rated eighteen of the children as having no deep attachments to anyone. In fact, eight of the children were described as emotionally detached. They were socially unresponsive and had no particular interest in anyone, but they exhibited few behavior problems. The remaining ten children were emotionally more animated and socially more engaging, but they were described as attention-seeking, clingy, and overly friendly with strangers, and they were more likely to exhibit problem behaviors. Interestingly, the eight children whom the nurses described as attached followed their preferred nurse around, protested her departures, and did not approach strangers.[30]

The eight children who were detached and unresponsive would now be described as showing signs of the emotionally withdrawn/ inhibited type of reactive attachment disorder. Since Tizard's landmark study, other research has demonstrated that this disorder is readily identifiable in some young children living in institutions and in young children when they are first placed in foster care following social neglect. By contrast, the disorder is rarely present in children adopted out of institutions.

In Tizard's study, the ten children who were more socially engaging but who had shallow emotions and connections to others today would be recognized as showing signs of indiscriminately social/ disinhibited reactive attachment disorder, or disinhibited attachment disorder, as it is also known. Subsequent research has demonstrated that this disorder also is discernable in socially neglected children in foster care. Unlike emotionally withdrawn/inhibited attachment disorder, indiscriminately social/disinhibited disorder is seen in a substantial minority of children adopted out of institutions. In fact, a continued high level of indiscriminate behavior is one of the most frequent social abnormalities in children adopted out of institutions, remaining evident for years after adoption in some cases.[31]

In BEIP, we assessed signs of the same kinds of behavior described in the Tizard study by interviewing caregivers and then having raters code the responses. We found that institutionalized children were rated by their caregivers as having significantly higher levels of both emotionally withdrawn/inhibited and indiscriminately social/disinhibited behaviors compared with the community children. Within the children living in institutions, however, there was no apparent relation between length of institutionalization and signs of either emotionally withdrawn/inhibited or indiscriminately social/disinhibited reactive attachment disorder.

For children living in institutions, the Carlson continuous ratings of the degree to which an attachment had formed were moderately correlated with caregiver ratings of emotionally withdrawn/inhibited attachment disorder, but they were unrelated to ratings of indiscriminately social/disinhibited attachment disorder. These findings suggest that emotionally withdrawn/inhibited RAD describes young children who lack attachment relationships with caregivers, whereas children may be attached or not and still show high levels of indiscriminate behavior.

Interrelationships among Attachment Measures in Children Living in Institutions

Among institutionalized children at baseline, continuous ratings of the degree to which attachment had developed were moderately correlated with caregiver ratings of emotionally withdrawn/inhibited RAD but unrelated to ratings of indiscriminately social/disinhibited RAD. This underscores the point that the emotionally withdrawn RAD describes the absence or near absence of formed attachments. High levels of indiscriminate behavior, however, were observed in children with secure, insecure, or disorganized attachments, and even in children with no attachments. This suggests that the indiscriminately social/disinhibited form of attachment disorder is largely orthogonal to classifications of attachment as measured by the Strange Situation.[32] Young children with attachment disorders are more likely to have disorganized or aberrant attachment patterns than organized patterns, but the two constructs of attachment disorders and patterns of attachment are distinctly different indexes.

Intervention Effects on Attachment

Placement in foster families was specifically designed to enhance the children's attachments and to reduce signs of disturbed attachment. To determine the degree of success, we examined both Strange Situation patterns of attachment and signs of attachment disorders.

Intervention Effects: Patterns of Attachment

At forty-two months of age, we re-administered the Strange Situation.[33] At this age, the procedure itself is quite similar to the procedure administered to toddlers, but the coding system is different. At forty-two months, assessing patterns of attachment rely not just on the child's movements toward and away from the caregiver but also on the child's use of language to regulate emotional distance and closeness. In preschoolers, the patterns are secure, avoidant, dependent, disorganized, controlling, and insecure/other. The last three are "atypical" and thought to be more associated with psychopathology than are secure, avoidant, and dependent patterns. Thus, the analyses we conducted were categorical comparisons of secure patterns versus all others, and typical (secure, avoidant, and dependent) versus atypical (disorganized, controlling, and insecure/other). We also used a continuous rating of security. As in the baseline assessments, the children who were living in institutions were seen with their favorite caregivers if they had a favorite and, if not, with someone who worked with the child regularly and knew the child well.

We predicted a higher proportion of secure attachments and higher ratings of security for children assigned to foster care versus those in care as usual. We also predicted fewer atypical patterns among the children in foster care.

In keeping with predictions, only 18 percent of the care-as-usual group was securely attached, but 49 percent of the foster care children and 65 percent of the community children were securely attached at forty-two months. On the continuous rating scale, community children were significantly more secure than the other two groups, but children in BEIP foster care were significantly more secure than children in the care-as-usual group.

We wanted to examine the effects of timing of foster care placement on security of attachment. In other words, was there an age at which the chances of a child's developing a secure attachment began to diminish? In fact, children placed in foster care before twenty-four months of age were significantly more likely to develop secure attachments than children placed after twenty-four months. Later placement did not preclude the child's developing a secure attachment to the foster parent, but it did reduce the likelihood of that happening. As shown in Figure 10.1, 25 percent of children older than twenty-four months at the time of placement managed to form secure attachments. *This finding is important in that it suggests that even children who have begun life in conditions of severe adversity have the capacity to recover substantially.* Moreover, as we will see in Chapter 11, children who had secure attachments at forty-two months were subsequently more successful socially and psychologically than those who had not.

Intervention Effects: Disorders of Attachments

Whereas the findings related to patterns of attachment were limited to forty-two months, findings on attachment disorders were available by caregiver report at every age and by observation at

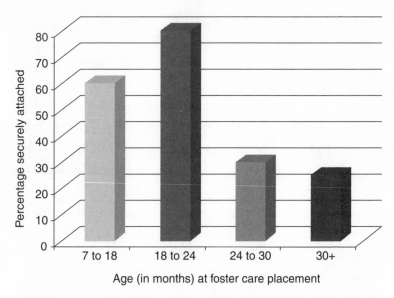

Figure 10.1. Percentage of foster care children with secure attachment relationships at 42 months as a function of age of placement. Note that significantly more children displayed secure attachments if they were placed at or
below 24 months of age.

fifty-four months and eight years. The effects of the intervention
on these outcomes were different for the two types of attachment
disorders.

For the emotionally withdrawn/inhibited type of RAD, significant positive effects of placement in foster care were evident at the
very first assessment age—thirty months—and at every subsequent
assessment. That is, the foster care children showed significantly
fewer signs of emotionally withdrawn/inhibited RAD compared
with the care-as-usual children, and the levels of this type of attachment disorder were indistinguishable from the levels in the community children. This means that all signs of disorder were effectively eliminated at the earliest possible age of assessment as a

function of the intervention. By thirty months of age, the children in foster care had received somewhere between one (the oldest children at the time the study began) and twenty-three (the youngest children at the time the study began) months of intervention, with an average of eight months. Nor were there timing effects— signs of emotionally withdrawn/inhibited reactive attachment disorder disappeared regardless of the age at which the child was placed in care. It appears that these behaviors are quite "plastic" and can be remediated with intervention.[34]

For the indiscriminately social/disinhibited type of attachment disorder, the response to the intervention was more complicated but also quite interesting. Caregiver and parent reports of indiscriminate behavior, as well as all data points from baseline through eight years, showed that children placed in BEIP foster care had fewer signs of indiscriminate behavior than did those receiving care as usual but more signs than children who had never been institutionalized. In addition, there were different recovery curves for indiscriminately social/disinhibited reactive attachment disorder in the foster care group and the care-as-usual groups, with decline of these signs in the foster care group evident between each pair of assessment times, with the lone exception of no change between forty-two and fifty-four months. In the care-as-usual group, differences were noted only between fifty-four months and eight years.[35] On the other hand, the assessment of overall group differences over time were only marginally statistically significant, indicating a much less robust response to intervention than for the indiscriminately social/disinhibited pattern.

We were also interested in supplementing caregiver reports with an observational measure of indiscriminate behavior. Anyone who has visited institutions for young children has noticed children who readily approach with arms upraised, as if you are a favored play-mate, even though these children have never seen you before. Such behavior does not occur in typically developing children because of so called stranger wariness, which typically appears by seven to nine months of age. Once stranger wariness becomes apparent, some degree of reticence with unfamiliar people persists for the next few years. In addition to our own frequent observations of this phenomenon, we were also aware that indiscriminate behavior was one of the most persistent and common abnormalities reported in children who had been adopted out of institutions in Romania and elsewhere.[36]

Initially, we considered implementing a procedure that would assess the children's willingness to "go off" with an unfamiliar re-search assistant who arrived at the lab. Because the children were used to visiting the lab and following research assistants to partici-pate in procedures, however, we were not sure that this would ac-curately reveal the behavior we were interested in describing. As an alternative, we decided to do something similar at the children's homes, meaning the apartments, houses, or institutions where they resided.

At fifty-four months of age, three home visits were scheduled. At the conclusion of the first home visit, the caregiver/mother was instructed about the second visit. A research assistant, unknown to the child, appeared at the door. Mother (or caregiver) and child were instructed to be present when the mother opened the door. The research assistant looked at the child and said, "Hello, my name is Florin, what is your name?" Then the research assistant, still look-ing at the child, said, "Come with me, I have something to show you." The mother/caregiver was instructed not to give any direc-tions to the child about what to do. We noted whether the child left with the stranger or refused to go. If the child did leave with the research assistant, the two walked a short distance, and then the child was greeted by a familiar research assistant from the week before, who said, "Hello, I have returned to play more games with you this week." This became known as the "Stranger at the Door Procedure."[37]

We believed that it would be unusual for four-year-old children to leave their homes to accompany a stranger, and the response of the community children suggested that we were correct. Only one of twenty-eight (4 percent) community children accompanied the stranger. In contrast, thirteen of thirty-one (42 percent) care-as-usual children accompanied the stranger. The foster care group was in between, with seven of twenty-nine children (24 percent) leaving with the stranger. The numbers did not include the full sample because we developed the procedure after the fifty-four-month assessments were already under way. By the time we implemented the procedure, only eighty-eight children actually completed it with usable data. A small number of parents declined or gave instructions to the child about what to do, and these were not included in data analysis. Although these group differences appeared striking, because the numbers of children included were small, there was actually no statistical difference between the proportion of care-as-usual and the foster care children who left with the stranger, though both groups were more likely to do so than were the community children. Importantly, there was a substantial agreement between caregiver reports of the children's behavior and observational codings from this procedure, indicating that they were tapping the same phenomenon.

With regard to timing of placement in foster care, results for indiscriminately social/disinhibited RAD were mixed. For children placed earlier (younger than twenty-four months), scores were generally lower, with the exception of forty-two months and eight years. For this early-placed group, scores dropped quickly between baseline and thirty months and then appeared to remain stable. For the later-placed group, a modest decrease in scores was noted between thirty and forty-two months, whereupon scores became stable again until the fifty-four-months–eight-year comparison, when they decreased.[38]

Demographic factors such as gender and ethnicity were unrelated to the recovery of young children from disorders of attachment. Children in the care-as-usual group who had cognitive delays showed less improvement in signs of emotionally withdrawn/inhibited RAD, but this was not true of the children in foster care.[39]

These findings indicate that placement in families clearly eliminated signs of emotionally withdrawn/inhibited RAD among young children raised in institutions, but it had a more attenuated effect on indiscriminately social/disinhibited RAD. This raises the question of what additional components should be included in interventions for indiscriminate behavior. For example, would training in reading and responding to social cues lead to reductions in indiscriminate behavior? Such training would be enhanced by better characterization of the social/cognitive abnormalities that presumably underlie indiscriminate behavior.

Relationship with Peers, Social Skills, and Social Competence

As children age, their social worlds expand from the immediate family to the school, classroom, and peers. The children in BEIP (even those who remained in the institutions over the years) at-

tended school, and in the institutions themselves, there were opportunities for social interaction. Thus we were interested in finding out about their emerging social skills and social competence
with same-age peers. To this end, we asked parents/caregivers to
report on the children's social behavior, and we also observed each
of the children in BEIP with a same-age, same-sex peer in our laboratory. We structured the interaction and videotaped it. We then
had these tapes coded for behaviors that were indicative of social
skills and competence.

Relatively little research has been conducted on the social behavior of institutionalized children among their peers, including their social skills and their close relationships and friendships.
Penny Roy and colleagues examined the social behavior of elementary school–aged children living in a residential care setting
compared with children reared in a foster family.[40] Using caregiver
reports, they found that one-fifth of the institutionalized children,
but none of the foster children, reported having few or no "selective" friendships. They concluded that this lack of social relationships was related to the children's lack of specific caregiver relationships. Similarly, Sandra Kaler and B.J. Freeman reported deficits in
the peer relationships of preschool-aged Romanian children living
in institutions when compared with a community sample.[41] Nese
Erol, Zeynep Simsek, and Kerim Munir examined adolescents' self-
reports of their social behavior in a sample of 11–18-year-olds living in institutional care in Turkey.[42] They found that, compared
with a community sample, caregivers and teachers of institution-
reared adolescents reported higher scores for adolescents' social
problems. Although these and other studies have reported social
difficulties among institution-reared children and adolescents, researchers did not examine children's social skills or specific behav-

iors in order to understand potential reasons for the lack of friendships and social difficulties in their samples. It is possible that children lack the skills necessary to form these types of relationships, as opposed to simply lacking a desire to engage in them in the first place.

It is important to understand the influences of institutionalization on children's social development throughout childhood for a number of reasons. Most important, early social behaviors set the stage for children's success at negotiating the increasing social demands and social changes that occur later in development. Social rejection or exclusion can lead to adolescents' involvement in delinquent or risk-taking behaviors.[43] Indeed, Erol and colleagues found that those institutionalized adolescents who reported fewer friends and poorer relations with friends also reported higher problem behavior scores on the Youth Self-Report.[44] Furthermore, difficulties with peer relationships in early childhood have been linked to difficult adjustment in adulthood.[45] To avoid such negative outcomes, it is important to determine whether foster care intervention can potentially remediate some of the negative social effects of previous institutional care.

Measurement of Social Development

To assess social skills, children's primary caregivers and teachers were asked to report on the children using the Social Skills Rating System (SSRS) when the children were eight years old.[46] The SSRS includes fifty-five items for caregivers and fifty-seven items for teachers that assess children's social skills (subscale scores for communication, cooperation, assertion, responsibility, empathy, engagement, and self-control), problem behaviors (subscale scores for externalizing, bullying, hyperactivity/inattention, internalizing, autism spectrum), and in the case of teacher-report only, academic competence (subscale scores for reading achievement, math achievement, and motivation to learn).

Children's interactions with an unfamiliar peer were assessed in the laboratory using six interactive tasks, presented in the following order: Tell Me, Lego, Puzzles, Jenga, Brainstorming, and Pacalici. In the Tell Me task children were asked to give each other a speech about themselves (two minutes each). In the Lego and Puzzles tasks, children were asked to work together, first to build a Lego structure and then to put together a set of puzzles. The Jenga task consisted of a block-tower building game in which players took turns removing a block from a tower and balancing it on top, but without knocking the tower down. In the Brainstorming task children were asked to discuss and list the top-three things they both like to do for fun. The Pacalici task consisted of a competitive Romanian card game. Each task lasted approximately five minutes.

Children's behavior was videotaped and subsequently coded independently by a team of Romanian research assistants. The behaviors coded included duration of speech, cooperation, off-task, on looking, conversation, fidgeting, social referencing, positive affect, and negative affect. We were particularly interested in the latency of beginning to talk during a particular session and the amount of give and take that a child would engage in during a particular task. The former set of behaviors we called speech reticence while the latter we called social competence.

Intervention Effects: Social Skills

We compared children's social skills as reported by caregivers and teachers in three groups: children who were randomly assigned to care as usual (CAUG), children who were randomly assigned to foster care (FCG), and children from the community (NIG). Results showed positive effects of the intervention on children's social behavior 3.5 years after the intervention ended.[47] Specifically, children in foster care were rated as more socially skilled by their caregivers (but not by their teachers) than were children assigned to institutional care. Interestingly, teachers rated children in both the FCG and the CAUG as having fewer social skills than did biological parents or foster caregivers.

Another important finding was a significant effect of age at placement into foster care on parent/caregiver ratings of social behavior. That is, children who were placed before twenty months of age were rated as more socially skilled by their teachers than were those placed later, even though teachers were unlikely to have known the age at which children were placed or even that they had resided in institutions. These results highlight the positive effect of early intervention on institutionalized children's social skills. However, foster care children were not rated as highly as the community sample, indicating a continued effect of early deprivation on development at age eight.

Intervention Effects: Peer Relatedness

We also found intervention effects when observing the children's behavior with peers.[48] Previously institutionalized children (CAUG or FCG) interacted with an age- and gender-matched unfamiliar peer from the community (NIG) on the tasks described above. We saw positive intervention effects; specifically, foster care children were rated as more competent and less reticent than institutionalized children. Children in the CAUG showed significantly higher levels of speech reticence, while children in the FCG showed significantly higher levels of speech competence. These results highlight the positive effects of intervention on the competence of children while interacting with unfamiliar peers in two socially demanding tasks.

An Actor–Partner Interdependence Model (APIM) was used to examine whether characteristics (for example, speech reticence) of ever-institutionalized children influence the behaviors of peers (in the areas of speech competence, social engagement, and social withdrawal) without institutional experience, and vice versa.[49] The speech reticence composite was chosen as the predictor because it was coded from the first interactive task (Tell Me) that the children engaged in during the assessment. With an APIM analysis we can see how one child affects the other's behavior in the dyad.

Not surprisingly, we found that when CAUG children displayed greater speech reticence, their NIG peers displayed less social engagement. However, there was no reciprocal influence of NIG children's display of speech reticence on CAUG children's social engagement. For the FCG-NIG dyads, there were no actor or partner main or interaction effects of speech reticence on social engagement during the Jenga/Pacalici tasks. Basically, FCG children had the social skills and competency to interact with unfamiliar peers so that they as a group did not affect the behavior of the community controls. By contrast, CAUG children negatively affected the behavior of the community controls.

Brain Behavioral Relations in Social Development

As described in Chapter 8, we found a timing effect for EEG alpha power at eight years, similar to the timing effect of age at placement on teacher-rated social skills at eight years. To explore the relations between brain functioning and social development, we decided to examine whether and how relational (that is, attachment security) and biological (alpha power) factors operate together to influence teacher-rated social skills in middle childhood.

We determined that children who were more securely attached to their caregivers at age forty-two months were more likely to be rated by their teachers as having better social skills at age eight. When we looked at alpha power at eight years, however, we demonstrated a statistical interaction, such that alpha power significantly moderated the relation between attachment security and social skills. For children with higher alpha power (defined as more than one standard deviation above the mean), greater attachment security significantly predicted better social skills, but for children with lower alpha power (defined as more than one standard deviation below the mean) there was no relation between attachment security and later social skills.

Thus differences in children's brain functioning (as indexed by alpha power) in middle childhood may reflect differences in early brain development resulting from early experiences with caregivers. Specifically, we can reasonably speculate that reductions in alpha power may reflect and influence attention- and sensory-processing abilities that may relate to social skills, including initiating contact with new peers, solving disputes, managing peer pressure, and regulating one's own emotions. It is hardly surprising that a secure attachment relationship during early childhood predicts a broad range of social skills years later. But this protective effect is seen primarily in those children whose EEG alpha power is more mature and age appropriate.[50]

Summary of Eight-Year Findings

This work revealed a number of remarkable findings (see Table 10.1). First, we continue to see intervention effects on social behav-

Table 10.1 Intervention and timing effects on socioemotional development

| | Assessment | | | | | | | | |
| | 42 months | | | 54 months | | | 8 years | | |
Construct evaluated	Intervention effects	Timing effects	Timing observed	Intervention effects	Timing effects	Timing observed	Intervention effects	Timing effects	Timing observed
Positive emotional expression	yes	no	—		not assessed			not assessed	
Attachment									
Security	yes	yes	24 mos.		not assessed			not assessed	
Organization	yes	yes	24 mos.		not assessed			not assessed	
Disordered attachment									
Inhibited	yes	no	—	yes	no	—	yes	no	—
Disinhibited	yes	yes	24 mos.	no	no	—	no	no	—
Social skills									
Teachers		not assessed			not assessed		yes	no	20 mos.
Parents		not assessed			not assessed			no	—
Peer relations									
Speech reticence		not assessed			not assessed		yes	no	—
Social withdrawal		not assessed			not assessed		no	no	—
Social engagement		not assessed			not assessed		no	no	—
Social competence		not assessed			not assessed		yes	no	—

ior and social competence eight years after the study began and four years after the intervention was completed. As in our other analyses, we used an intent-to-treat approach so that the positive intervention effects that we found are actually a conservative estimate of the true effects. Second, timing is important for the intervention effects, with children taken out of the institution before two years of age displaying better social skills than those placed into our intervention after age two. It is also clear that there are lasting effects of exposure to social deprivation among the sample. As a group the children at age eight still display lower social skills compared with typical community children. Finally, linking brain functioning to social development further underscores the importance of early caregiving experiences on subsequent development.

Conclusion

This chapter highlights findings about emotional responsiveness, attachment, and social development in middle childhood and represents some of our most powerful and compelling findings in the Bucharest Early Intervention Project. The effects of institutional rearing in these domains are quite profound—but so is the children's capacity to recover. Some of the earliest and fullest recovery we have seen among formerly institutionalized children placed in foster care are in the realms of their social and emotional functioning. For example, in the puppets and peek-a-boo tasks, as early as the very first assessment at thirty months, the positive emotional responsiveness of children in foster care was comparable to that of children who had never been institutionalized and far better than that of the children who remained in institutions—full recovery on this measure. The normalization of signs of reactive attachment disorder by thirty months was similarly evident. Further, gains in attachment security for children continued to be profound into the fourth year of life and to predict a wide range of subsequent outcomes in the domains of psychopathology and social development. Developing a secure attachment to a foster parent was centrally involved in the mechanism or process through which these young children—who began their lives in conditions of serious adversity—changed to more adaptive developmental trajectories. Finally, the gains in middle childhood, years after the intervention ended, are associated with functional brain changes.

These findings also indicate that the social and emotional domains of development have considerable plasticity, seemingly even more than in the cognitive realm (see Chapter 7). In other words, though the children's social and emotional functioning was adversely affected while they lived in institutions, once placed in more normative caregiving environments, they demonstrated remarkable recovery. But it is also clear that the sooner environmental remediation occurs the better, as reflected in repeated demonstrations of sensitive periods in these same domains. In the final chapter of the book, we return to these issues as we consider BEIP contributions beyond the children we studied.

In the future we hope to address the effects of the transition to adolescence—when the complexity of social demands and the prevalence of emotional problems typically increase—in children with histories of severe deprivation. We anticipate significant challenges for the children, and we expect their concurrent environments to have important effects. Nevertheless, it remains to be seen if the very early experiences we have documented are still detectable years after the intervention.

Chapter 11

Psychopathology

Thus we reached the conclusion that loss of mother-figure, either by itself or in combination with other variables yet to be clearly identified, is capable of generating responses and processes that are of the greatest interest to psychopathology.

John Bowlby, *Attachment* (1969)

In April 2010, a Tennessee mother sent her seven-year-old adopted son alone on a flight to Moscow, with a note telling Russian authorities that she no longer wished to parent the boy, whom she had adopted six months earlier. She complained that he had severe psychological problems, including psychopathic tendencies and violent behavior. The orphanage he came from, she claimed, had lied about his condition and the nature of his problems.[1] For these reasons, she had decided to relinquish the adoption and, amazingly, to return the boy to Russia.

Though clearly this is an extreme case, according to ABC News, fifteen children adopted from Russian institutions were murdered by their adoptive parents in the United States over a fifteen-year period beginning in the early 1990s.[2] In these cases, parents perceived the children to be unmanageable and violent, although more than half of them were less than three years old.[3]

Such acts are rare, of course, but as we've established, many children adopted out of institutions exhibit a variety of developmental, behavioral, and emotional problems. Nearly fifty years ago, John Bowlby, in a discussion of psychological problems among institutionalized children, highlighted "a tendency to make excessive demands on others, and to be anxious and angry when they are not met," as well as "a blockage in the capacity to make deep relationships, such as is present in affectionless and psychopathic personalities."[4] In the past fifteen years, adverse experiences associated with institutional rearing have been increasingly linked to serious psychopathology. Adoption studies have demonstrated clearly that children raised in institutions experience elevated rates of certain types of psychopathology. For example, Rutter and the English and Romanian Adoptees Study group have emphasized disinhibited attachment disorder, inattention/overactivity, quasi-autism, and intellectual disabilities as a post-institutional syndrome.[5] Other studies had described attachment disorders, stereotypies, and internalizing and externalizing behavior problems as common sequelae of institutional rearing.[6]

When BEIP began, there had been little systematic research on psychopathological outcomes associated with institutional rearing. The extant work primarily focused on internationally adopted children who had begun life in an institution. These studies did not assess children in the institutions before adoption, making it unclear when their problems emerged. Moreover, as noted, adoption studies are by nature unrepresentative, since only certain children are selected for adoption.

Previous reports of psychiatric disorders in children with histories of institutional rearing typically focused only on a single disorder.[7] Other than ADHD, disruptive behavior disorders had not received careful study, nor had anxiety disorders beyond post-traumatic stress disorder (PTSD). Other studies were limited to parent checklist reports of behavior problems without systematic diagnostic interviews.[8] Precursors of psychopathology had received little attention, and the mechanisms through which psychopathology emerges in children reared in institutions had not been addressed. We considered all these gaps in our understanding of the psychopathology associated with institutional rearing when designing the BEIP.

Baseline Assessments of Infants and Toddlers

Recall that when we implemented the study, children were six to thirty months of age, and the study group at baseline comprised 136 children who were living in institutions and had lived in institutions at least half their lives (see Chapter 2).

Because of the very young age of the children, we used a parent/caregiver report measure, the Infant Toddler Social Emotional Assessment (ITSEA), to identify behaviors that might indicate or be precursors of psychopathology.[9] The ITSEA is a 195-item questionnaire, which caregivers/parents in the study used to rate problem behaviors and competencies in children twelve months and older. *Externalizing problems* (overactivity/impulsivity, aggression/defiance, peer aggression), *internalizing problems* (depression/withdrawal, anxiety, separation distress, inhibition to novelty), *dysregulation* (negative emotionality, sleep and eating problems, sensory sensitivity), and overall *competence* (compliance, imitation/play, attention, mastery motivation, empathy, prosocial peer relations) scores were computed from sums of individual item ratings. Scores also were derived for *maladaptive behaviors, social relatedness,* and *atypical behaviors.* Because the ITSEA had been normed previously by age group and gender, a given child's score could be compared with scores for children of similar age and gender.[10]

Institutionalized and Community Children at Baseline

At baseline, caregivers reported that children reared in institutions had more maladaptive and more atypical behaviors than children from the community group. Surprisingly, there were no overall differences between groups on the externalizing, internalizing, and dysregulation scales. By contrast, children raised in institutions were rated lower on both the competence scale and the social-relatedness scale than were community children.[11]

The lack of differences between institutionalized and never-institutionalized community children in internalizing and external-izing was unexpected, in that most other measures had shown large differences between the two groups. Perhaps the children were too young (an average of twenty-two months at baseline) to display behaviors that might be indicative of or precursors for psychopathology, or at least too young for their caregivers to recognize these behaviors.

Intervention Effects in Preschool Children

When children were fifty-four months of age, we were able to use a structured psychiatric interview, the Preschool Age Psychiatric Assessment (PAPA), in place of checklists.[12] This assessment had been used previously with children aged two to five in a large study of children during primary care visits in North Carolina, demonstrating test-retest reliability comparable to that of structured psychiatric interviews used to assess older children and adults.[13]

We also used the PAPA later in Bucharest in a study of the prevalence of psychopathology in young children visiting pediatric clinics.[14] As shown in Table 11.1, the rates of disorders in our community sample at fifty-four months were generally higher than those of this large sample of two- to five-year-old children recruited a few years later from pediatric clinics in Bucharest.

We took several steps to ensure that the psychopathology and impairment data were collected properly. The lead interviewer was trained in administration of the PAPA by the group at Duke University who developed the measure, and he subsequently trained other Romanian interviewers. After the PAPA was translated into Romanian, it was back translated into English and assessed for meaning at each step by bilingual research staff.

Caregivers and parents were interviewed in detail about signs of psychiatric disorder in each child, as well as about the degree to which these signs impaired the child's school and social functioning. For children living with biological parents or foster parents, the mother reported on the child's behavior. For children living in institutions, an institutional caregiver reported on the child's behavior. As with other measures, the caregiver chosen was the one considered the child's favorite, according to staff consensus. If the child had no favorite, we used a caregiver who worked with the child regularly and was said to know the child well. Detailed questions about the frequency, duration, and dates of onset of symptoms of a large number of psychiatric disorders were administered. Algorithms were then applied to responses to determine whether each child met diagnostic criteria for specific disorders. Further, caregivers and parents were interviewed about the degree to which symptoms interfered with the child's functioning at home, at school, and in public. For purposes of the study, we derived diagnoses to

Table 11.1 Rates of psychiatric disorders in 2–5-year-olds in Bucharest epidemiological study, the BEIP community group, and the BEIP ever-institutionalized group at 54 months

Psychiatric disorders	Bucharest epidemiological sample $n = 1003$	BEIP community sample (never-institutionalized group) $n = 72$	BEIP ever-institutionalized group $n = 111$
	Prevalence rates (percentage)		
Major depressive disorder	0.2	0.0	2.7
Any anxiety disorder	4.4	13.6	30.6
Post-traumatic stress disorder	0.2	0.0	0.0
Attention deficit hyperactivity disorder	0.4	3.4	20.7
Reactive attachment disorder (inhibited)	0.0	0.0	4.1
Reactive attachment disorder (disinhibited)	2.0	0.0	17.6
Any emotional disorder	5.4	13.6	32.4
Any behavioral disorder	1.4	6.8	27.0
Any psychiatric disorder	10.5	22.0	53.2

examine prevalence of psychiatric disorders, symptom counts for constructs such as internalizing, externalizing, inattention, and aggression, and degree of functional impairment.

History of Institutional Rearing

We found that the majority of four-and-a-half-year-old children with a history of institutional rearing had diagnosable psychiatric disorders (53 percent), whereas never-institutionalized children from the community were significantly less likely to have a disorder (22 percent). In addition, 52 percent of the children with a history of institutional rearing had two or more psychiatric diagnoses, whereas only 20 percent of the community children had multiple diagnoses. This finding suggests that institutional rearing is an extraordinarily powerful risk factor for psychopathology. Table 11.1 provides the data comparing the children who had a history of institutionalization, the community control sample, and the larger community sample that we had collected in Bucharest.

Intervention Effects

To examine the effects of placement in foster care, we looked at the prevalence of disorders in children in FCG versus those in CAUG. Children in foster care were significantly less likely to have a disorder (46 percent) than those who received care as usual (62 percent). Although there were no significant differences in externalizing disorders between the groups, there were clear group differences in internalizing disorders (22 percent in FCG versus 44 percent in CAUG). So although the intervention reduced psychopathology, the differences between the groups were explained entirely by the differences in internalizing disorders.

Unfortunately, few comparable intervention studies exist that might help us understand this selective effect of foster placement on reducing internalizing rather than externalizing disorders in preschool children. One study we did identify, however, found that preschool children in foster care who had been exposed to drugs prenatally or who had lost multiple primary caregivers seemed to demonstrate consistently greater levels of aggression and had more difficulty developing trusting connections with their caregivers, even when intensive behavioral treatment was provided to foster parents.[15] Many children in BEIP likely shared the same two risk factors as the maltreated preschool children—prenatal exposure to drugs and alcohol and exposure to psychosocial deprivation early in life—and these may have contributed to failure of the intervention to reduce signs of externalizing disorders.

Moderators of the Intervention

In psychological research, a moderator is a variable that affects the direction and/or strength of the relation between an independent or predictor variable and a dependent or criterion variable. In randomized clinical trials, a moderator variable specifies for whom a treatment works and under what conditions.[16] Virtually all interventions are effective only for some participants, so an important question is, who responds and who does not? Below we describe three potential moderators—gender; age at which the child was placed in foster care; and individual variations in specific genes that may moderate the effects of early experience—as variables that we thought might help define who responded to the intervention with fewer signs of psychopathology.

Gender

At fifty-four months, boys were more symptomatic than girls, regardless of their caregiving environment. Moreover, unlike girls, boys had no reduction in total signs of psychiatric disorders following foster placement. These findings were surprising because until this point we had identified few results that differed by gender, and in the only other study assessing psychopathology in preschool children using the PAPA, no gender differences were found.[17] An initial question for us was how to understand why boys did not benefit from foster care placement when it came to signs of psychopathology.

We considered a number of possible explanations for this finding. First, boys are known to be more vulnerable biologically than girls; for example, the prevalence of pre- and perinatal complications tends to be higher in boys than in girls, and boys suffer from more neurodevelopmental disorders (for example, autism) than girls.[18] Second, boys are less mature than girls in the early years, and therefore the abilities needed for adaptive behavior regulation may be underdeveloped, increasing their vulnerability in deprived environments. Indeed, this immaturity continues through the adolescence years, when boys' brain development tends to lag behind that of girls by one to two years. Even among the community children boys had more symptoms, more disorders, and more impairment than did girls.[19] This suggests that the institutional rearing environment may exacerbate a difference that is already evident.

Third, it is possible that caregivers "preferred" girls to boys. If so, might they have responded differently to externalizing behaviors in boys and girls? While we found no evidence of interactive differences of foster parents with boys and girls when we examined them in unstructured activities, it may be that boys required different experiences with caregivers than did girls to develop better emotion-regulation strategies. As we will see later in this chapter, boys were less likely than girls to form secure attachment relationships with their foster caregivers, which in turn mediated the relations between institutionalization and disorder.

Timing of Placement in Foster Care

A central question in BEIP has been whether timing of the foster placement was related to intervention effects. As noted, for IQ, EEG, security of attachment, language skills, and other outcomes, earlier placement was associated with more favorable outcomes. In the case of psychiatric symptoms, disorders, and impairment, there were no timing effects. Entering foster care mattered greatly, but whether placement occurred at seven months, when the youngest child was placed, or thirty-three months, when the oldest child was placed, did not matter. This result was surprising to us; we would have expected to identify a sensitive period before which intervention might exert an influence on behavioral outcomes.

There are a number of possible reasons for the lack of a finding here. First, the youngest infants in our study were six months old at the start of the intervention. The sensitive period for behavioral outcomes of internalizing and externalizing may be younger than that age, and so we may have missed identifying this period. Second, the effects we noted may be a function of prenatal exposure to alcohol or genetics of the biological family, making them less susceptible to intervention than other domains. As we will see in the following section, individual differences in the child's genetic make-up may play a role in differential intervention outcomes.

Genetic Polymorphisms

A burgeoning area of research in developmental psychopathology that offers the potential to better define individual differences in vulnerability, resilience, and recovery centers on the impact of common, but functional, differences in the DNA sequence of specific genes.

A Brief Tutorial on Genetics

The human genome is the complete set of genetic information of an individual, an exact copy of which is found within almost every cell in the body. An individual's genome is recorded in long strands of deoxyribose nucleic acid—DNA—the code that tells each cell exactly what to do, which proteins to make, and when to make them. The individual differences in this sequence, or genetic code, explain, in part, why we are different from other humans, and different from animals. These strands, together with proteins, constitute chromosomes. Humans have twenty-three pairs of chromosomes, one from each parent, totaling forty-six chromosomes, found in almost every human cell, the exception being sperm, oocytes, and mature red blood cells.

DNA, within the chromosomes, is arrayed as a double-helix molecule consisting of a chain of linked units called nucleotides. Each nucleotide consists of a sugar (deoxyribose), a nitrogenous base, and a phosphate group that together provide the structure or "backbone" of the DNA double helix. The two strands of DNA are joined by electron bonds formed by specific pairs of nucleotides, constituting the actual genetic information of the individual. In DNA there are two purine bases, adenine and guanine, and two pyrimidine bases, cytosine and thymine (ACGT). Adenine is always paired with thymine and guanine with cytosine in the complementary strand of DNA. Genes are functional units of inheritance located on chromosomes that are defined sequences of the nucleotides. The specific sequence of nucleotides is the code that is needed to make specific ribonucleic acid (RNA) messengers, which then leave the center of the cell, the nucleus, and travel to other parts of the cells to make proteins.

Humans have an estimated 20,000 genes. The specific sequence of each gene is found in the same place on each chromosome. Genes influence our traits and tendencies through the proteins they create. Protein synthesis is a two–stage process. In the first, transcription, one strand of DNA serves as a template to create a molecule called messenger RNA (mRNA). In the second stage, translation, the mRNA serves as a template to create a protein. Although some individual traits, or phenotypes, may be caused by single genes, the majority of personal characteristics, such as height, are likely influenced by a combination of genes.

A gene is a basic sequence of DNA, while an allele is one of the two forms of a gene that an individual gets from each parent; it may or may not have small differences in the actual sequence of the gene. Humans have one copy of each gene (and therefore one allele) on each chromosome. If both alleles are the same, they are homozygotes; if different they are heterozygotes. Sometimes, different alleles can result in different observable phenotypic traits, but most variations, at the level of the DNA sequence, result in little or no observable variation. Therefore all people carry the same genes, but certain people may have a specific allele of that gene, and this minor sequence difference can result in differences in a particular trait. These sequence differences, when common enough in a population, are called polymorphisms. They are what is inherited between parents and children, and in part explain why children resemble their parents. Multiple types of sequence variations exist, and though the majority of polymorphisms are not expected to alter the function or regulation of a gene, certain polymorphic variants are associated with functional changes, which can influence the function of the gene to varying degrees.

One commonly used polymorphism is called a single nucleotide polymorphism (SNP). An SNP is a variance in a single nucleotide—A (adenine), T (thymine), C (cytosine), or G (guanine)—that differs between individuals. To illustrate, imagine that a DNA sequence in one individual is ATTGCCA and in another is ATTG-GCA. The two sequences differ by a single nucleotide, a guanine for cytosine substitution. Since genes occur in pairs (one from each chromosome an individual inherits from his or her parents), this substitution may occur in one or both alleles of that particular gene.

Large maps of SNPs across the genome are now known and used extensively to define large areas of the genome as well as specific genes associated with particular outcomes. SNPs can also be thought of as street signs, a way to mark or label a particular area of the genome. Not all SNPs are functional, and each gene can have more than one SNP, as well as other forms of polymorphisms. Although even nonfunctional SNPs can be used to suggest an association between a gene and a particular phenotype, because they suggest that phenotype is associated with the specific area of the genome where that gene resides, SNPs that alter the function of the gene provide enhanced evidence for a causative relationship. Similarly, other polymorphisms, such as variable number tandem repeats (VNTR), which are repeated blocks of a particular sequence of nucleotides where the number of the blocks, as opposed to the actual sequence itself, is what varies between individuals, may be functional or nonfunctional.

While the respective roles of genes (nature) versus experience (nurture) was once intensely debated, there is now broad agreement that both genes and environment contribute risk in development and psychopathology. Increasingly, the focus is more specifically on the interaction between genes and the environment and how this interaction may influence an individual's risk and resilience.[20]

Critical to BEIP is an exploration of "gene-by-environment" interactions, where there is a demonstrable interaction between functional polymorphisms and a specific aspect of the environment. In psychiatric genetics research, gene-by-environment studies are increasingly being used to define who succumbs to the effects of adverse environmental conditions and who does not. A genetic *diathesis-stress model* of psychopathology, for example, posits that genetic polymorphisms confer risk, but that psychopathology results only if the susceptible individual experiences specific or nonspecific stressors.

A number of gene-by-environment interaction studies have now demonstrated that the interaction between adverse early experiences and individual differences in genetic polymorphisms results in variability in outcomes.[21] For example, variants of genes are involved in regulating how much of a neurotransmitter such as serotonin or dopamine is available in various circuits in the brain.

Over the past few years we have grown increasingly interested in the role of individual differences in genetic variation and how these differences interact with a child's experience to yield particular patterns of behavior and psychopathology. The RCT design of the BEIP permits a more precise examination of gene-by-environment interactions, meaning that variations in environments in combination with subtle genetic differences may yield different outcomes. This research design thus permits the systematic evaluation of the contribution of genetic variation to different types of psychopathology.

In BEIP, two moderation questions can be addressed. First, does genetic variation confer increased vulnerability to the development of psychopathology? That is, do specific polymorphisms confer susceptibility to specific types of psychopathology, depending on the child's rearing environment? Second, how does the individual child, as a function of genotype, respond to environmental enhancement? These effects of increasing or decreasing risk for psychopathology are presumed to be the result of the functional differences in the gene product (that is, protein) conferred by the sequence variation.

Explorations of genetic contributions to psychopathology are interesting and important, but some of the initial enthusiasm in this area has waned because the literature is replete with failures to replicate what appear to be exciting findings. Different studies have yielded contradictory results in many areas. The problem seems to be that the opportunities for chance findings in these analyses are considerable (in part owing to small or biased samples). Therefore, neurobiologically sound hypotheses about putative genetic contributions to variability in outcomes are key to this exploration. Stacy Drury, a child psychiatrist and molecular geneticist at Tulane University, has led our exploration of genetic factors in BEIP. One approach has been to examine polymorphisms in genes that have been previously demonstrated to be involved in specific neural circuitry important for the outcomes being investigated. The first area we explored was depressive symptomatology.

Depression

The catechol-O-methyltransferase (COMT) gene is located on chromosome 22q11, and the gene product is expressed especially in the hippocampus and the prefrontal cortex. COMT is an enzyme that is the main regulator of degradation (breakdown) of neurotransmitters such as dopamine and other catecholamines in the prefrontal cortex.[22] Multiple polymorphisms within the COMT gene impact levels and functional effects of the enzyme.

Neurotransmitters

Neurotransmitters are chemicals that are transmitted from one neuron to the next across synapses (the site where nerves connect to one another). Neurotransmitters are released from the axon of one neuron into synaptic space in response to electrical impulses that travel through the nerve cell. After release into the synaptic space, the neurotransmitter is taken up by receptors in the dendrite or cell body of another neuron.

There are probably more than 100 neurotransmitters in the human brain, but catecholamines (dopamine, epinephrine, and norepinehrine), serotonin, glutamate, and GABA (gamma-aminobutyric acid) are among the most common and important in psychiatric studies. Different circuits in the brain are rich in one or more neurotransmitters.

The most frequently studied polymorphism in the COMT gene is the val[158] or [108]met amino acid substitution. An individual can be met/met, val/val or met/val in the pair of alleles of the COMT gene. The protein produced by the met allele variant has a fourfold decrease in COMT enzyme activity, resulting in slower degradation of dopamine in the prefrontal cortex. Medications that prolong dopamine levels are used to treat depression, so it made sense to look at polymorphisms affecting levels of dopamine when examining vulnerability to and protection from depression in adverse environments. Because the met allele reduces enzyme activity, we would expect higher levels of dopamine in individuals with this allelic variation and therefore protection from depressive symptoms. In fact, previous research generally had demonstrated that individuals who had the COMT [108]met variant were less likely to become depressed, though there were some inconsistencies in the studies. Therefore, we hypothesized that the COMT [108]met variant would protect against the development of depressive symptoms in BEIP children. We also assumed that the impact of COMT variation would differ between the CAUG and the FCG, such that with more prolonged exposure to the negative and stressful environment of institutions, the CAUG would demonstrate a larger impact of the genetic variation.

We explored the effect of different versions of the COMT val-[158]met on depressive symptoms in two different ways. First, we compared individuals with any copies of the met allele (that is, met/met and met/val) with homozygote val/val individuals. We found that signs of depression were significantly lower among children with at least one copy of the met allele, even after adjusting for group and gender. That is, the met allele was protective only for CAUG children but not for FCG children. This means that genetic differences, in the form of allelic variations, had effects, but only for children who had been exposed to the more extreme adverse environment.[23] For children placed in foster care, who we had demonstrated had responded to the intervention with fewer signs of depressive disorders at fifty-four months than had children in the CAUG, the protective effect of the met allele of COMT was not evident.

Second, we explored whether there was a "dose-dependent" effect within the CAUG related to the number of met alleles. We found that, in fact, this was the case: children with met/met were more protected than those with met/val, who were more protected than those with val/val.

These results were the first report of a gene-by-environment interaction in a setting of early, severe social deprivation. We demonstrated that vulnerability to signs of depression in young children resides within both individual characteristics and adverse environments. That is, the highest level of depressive signs were seen in children with the val/val allelic variation of COMT and more prolonged exposure to the adverse environment of institutional rearing.[24]

Indiscriminate Behavior

We next sought to examine whether genetic polymorphisms could explain some of the variation in children with regard to indiscriminate behavior, one of the most common abnormalities reported in young children raised in institutions. Specifically, we examined whether the interaction between functional polymorphisms in Brain-Derived Neurotrophic Factor (BDNF) and the Serotonin Transporter gene (5HTT) interacted with group placement to predict indiscriminate behavior.

Here, the question was somewhat different from what it had been in the COMT analysis. Jay Belsky and colleagues described a *differential susceptibility model* of genetic contribution to psychopathology and adaptation.[25] In this model, specific polymorphisms confer a differential responsiveness to the environment, instead of risk or resilience per se. So-called plasticity alleles or responsivity alleles are believed not only to enhance outcomes in positive environments but also to increase vulnerability in adverse environments. Individuals with a less sensitive or responsive allelic variant are expected to demonstrate much more limited variability in outcomes (that is, exhibit less sensitivity).

When we examined indiscriminate behavior as an outcome, which was less responsive than other psychopathologies to the foster care placement intervention, we found cumulative differential susceptibility in children defined by the presence of plasticity alleles—the s/s genotype of the serotonin transporter gene or met carriers of the val[66]met polymorphism of the brain-derived neurotropic factor gene. Specifically, the more responsive alleles children had, the greater the difference in indiscriminate behavior between the CAUG and the FCG.

At baseline, before any alterations in the environment, we found no demonstrable genetic influence on indiscriminate behavior. However, over the course of the intervention, a clear pattern developed. Specifically, children with more "sensitive" alleles who remained in the institutional setting continued to have the highest levels of indiscriminate behavior (worst outcomes), whereas children with those same "responsive" alleles who had been randomized to the foster care intervention demonstrated significant decreases in indiscriminate behavior (best outcomes). Children with either the long allele of the 5HTT or the val/val genotype of BDNF demonstrated little difference in levels of indiscriminate behaviors over time and no group-by-genotype interaction.[26] These findings represent the first genetic associations reported regarding indiscriminate behavior, and they add to the growing body of literature supporting a differential susceptibility model for early development.

Externalizing Disorders

Signs of externalizing disorders showed no intervention effect at fifty-four months. This finding made it possible for us to examine the differential susceptibility model using a neurobiological-based genoset (combination of genes) that leveraged known previous associations with externalizing behaviors and evidence of molecular interaction between the following genotypes: the dopamine transporter gene VNTR polymorphic variants (DAT 3'UTR VNTR), the serotonin transporter gene polymorphic variants (5HTT s/l), and the brain-derived neurotropic factor polymorphic variants (BDNF val^{66}met). We constructed a categorical cumulative sensitivity genotype based on previous studies, having established that there were molecular and biological interactions among these genes, and that they were all relevant to stress-related psychopathology. Controlling for gender and ethnicity, we tested the association between cumulative susceptibility genotype and externalizing behavior, using the Preschool Age Psychiatric Assessment at fifty-four months.

At fifty-four months of age, children with three susceptibility genotypes (10/10 homozygous for the DAT 3'VNTR 10 repeat polymorphism, s/s homozygous for the 5HTTLPR polymorphism, and BDNF Met carriers of the val^{66}met polymorphism) had both the greatest number of externalizing signs in the CAUG and the *fewest* signs in the FCG. Children with zero or one of the susceptibility genotypes did not differ between the two groups in signs of externalizing behavior.[27]

These findings offer significant support for the theory of genetic differential susceptibility. A subset of children, defined by cumulative genetic factors, are the most vulnerable to adverse caregiving environments but also the most able to recover if they are placed in high-quality caregiving environments.

Mediators of the Intervention

Another important issue in intervention studies is the mechanism or means through which the positive outcomes are achieved. Factors that may explain mechanisms of interventions are called mediating variables.[28] Mediation is a statistical analysis that allows us to examine a pathway between a risk factor, in this case foster placement versus care as usual, and an outcome, in this case internalizing disorders. The question is what plausible pathway could explain why placement in foster care in the first thirty months of life would lead to a reduction in anxiety and depressive symptoms at fifty-four months. Here, we drew upon what we knew from our previous findings in BEIP and what the literature from other studies suggested about potential factors associated with psychopathology.

Kate McLaughlin, a clinical psychologist and epidemiologist at the University of Washington, examined security of attachment as a potential mediator of the intervention on internalizing disorders. Her reasoning was that since a young child develops a secure attachment relationship with a primary caregiver when the caregiving relationship is predictable, sensitive, and responsive to the child's needs, and since our foster care intervention explicitly encouraged foster parents to form committed attachments to the children in their care, it made sense to examine whether security of attachment related to reduction in psychopathology.

We already had demonstrated that attachment was severely compromised in children reared in institutions (see Chapter 10). We also knew that in some other high-risk samples, insecure and especially disorganized attachment was associated with risk for internalizing psychopathology in children, including anxiety disorders and major depression.[29]

We knew that girls in the FCG had fewer internalizing disorders than girls in CAUG. We also knew that the intervention had no effect on internalizing disorders in boys.[30] Finally, at forty-two months, we knew that girls in the FCG were more likely to have secure attachments than girls in the CAUG, but we observed no difference in boys.[31] McLaughlin then showed that greater attachment security predicted lower rates of internalizing disorders in both sexes.[32]

Thus the conditions necessary to test for mediation were met. That is, FCG girls had fewer signs of internalizing disorders than CAUG children at fifty-four months of age. FCG children had more secure attachments at forty-two months than did CAUG children. Finally, secure attachment at forty-two months predicted signs of internalizing disorders at fifty-four months. This finding allowed a test of security of attachment as a mediator, as shown in Figure 11.1.

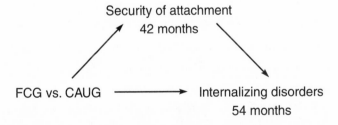

Figure 11.1. Security of attachment mediates the effect of foster care placement on reduction of internalizing disorders.

With security of attachment included, the direct pathway between group placement and internalizing disorders was no longer significant. This meant that development of attachment security at forty-two months fully mediated the intervention effects on internalizing disorders in girls. In other words, the reason foster care reduced internalizing disorders in girls was that they formed secure attachments. These additional analyses also clarified that greater attachment security at forty-two months predicted lower rates of internalizing disorders in *both* girls and boys. Thus the reason boys did not show reductions in internalizing disorders is that they were significantly less likely than girls to form secure attachments at forty-two months. These results provide compelling evidence for the crucial role of constructing secure attachments in the reduction of internalizing disorders in young children exposed to institutional rearing.

Lucy McGoron, a developmental psychologist working at Tulane University, conducted similar mediational analyses, examining security of attachment as a mediator of the effect of observed caregiving quality at thirty months on psychopathological outcomes at fifty-four months, including signs of internalizing disorders, externalizing disorders, RAD, indiscriminate behavior, and stereotypies, as well as overall psychiatric impairments. For each type of disorder, security of attachment mediated the relations between caregiving quality and signs of the disorder. For these analyses, McGoron used the entire group of FCG and CAUG children, so rather than a test of mechanisms of the intervention, these analyses highlighted an important pathway through which deprived children may be more or less likely to develop psychopathology. These findings have implications for young children with histories of institutional rearing, in that forming secure attachments seems an especially important means of reducing the likelihood that they will develop subsequent psychopathology.[33]

Brain Structure, Function, and Psychopathology

One of our interests from the outset was exploring brain behavior relationships, including brain functioning in different types of psychopathology. Preliminary explorations in this area have yielded some promising findings.

ADHD and Brain Structure and Functioning

ADHD is a highly heritable neurodevelopmental disorder that is extraordinarily prevalent in children with histories of institutional rearing.[34] In BEIP, for example, 21 percent of the FCG and 23 percent of the CAUG were diagnosed with ADHD when they were fifty-four months of age. Since most abandoned children in Romania were placed in institutions at birth, and since ADHD is not detectable in early infancy, it is unlikely that parents would have noted some characteristic of ADHD in their newborn and on that basis abandoned the child to an institution. This suggests an alternative pathway other than the usual genetic vulnerability to ADHD seen in samples of more typically reared children.

There are at least two possible mechanisms for the high prevalence of ADHD in institutionally reared children. First, some aspect of prenatal experience (for example, prenatal alcohol exposure) may have contributed to the development of the disorder. Second, the deprived environment of institutional rearing may somehow be implicated in the etiology of ADHD. This is a provocative hypothesis, because the hereditability of ADHD is so well-established. On the other hand, hereditability estimates are population specific, and the hereditability estimates for ADHD derive from samples quite different from children raised in institutions.[35] For reasons we have discussed, we have little information about the details of prenatal experience among the children in BEIP, so it is difficult for us to determine which of these rival hypotheses explains the association between institutional rearing and ADHD.

One possible line of inquiry is to determine if children in BEIP have structural brain differences that have been previously described in clinical samples of children with ADHD. To pursue this question, Kate McLaughlin conducted analyses of the MRI scans we collected when the children were eight to ten years of age.[36]

To target our analyses, she noted that there are two neurodevelopmental pathways that have been implicated in ADHD in noninstitutionalized children. One pathway connects the prefrontal cortex (PFC) and the striatum. Considerable evidence has demonstrated both smaller PFC and striatal volumes as well as abnormal activation and functional connectivity between these regions, particularly during executive functioning tasks.[37] In fact, the assumption has been that abnormal fronto-striatal circuitry is responsible for ADHD behaviors. More recently, however, some evidence suggesting that ADHD may involve widespread reductions in cortical thickness across not only in the PFC, but also in the parietal and temporal cortices.[38]

We found that children exposed to institutional rearing had reduced cortical thickness in the PFC as well as parietal and temporal cortices—findings quite similar to the pattern of widespread cortical thinning observed in children with ADHD.[39] Also, we found that institutional rearing was unrelated to volume of the striatum.

Because previous research had suggested important structural and functional differences in the striatum, particularly the caudate, among children with ADHD compared with healthy controls, we suggested that striatal contributions to ADHD symptomatology may reflect primarily genetic and prenatal influences whereas cortical mechanisms may include postnatal influences. This speculation will have to be pursued in future research.

We conducted additional analyses to examine brain structure mediating institutional rearing on children's behavior. Although reductions in cortical thickness were widespread, those in the prefrontal and parietal cortex explained elevations in inattention, and reductions in thickness in the prefrontal, parietal, and temporal cortex explained elevations in impulsivity observed in these children. Early-life psychosocial deprivation seems to have disrupted cortical development, culminating in heightened risk of ADHD.

If ADHD involves fundamental deficits in the ability to both generate accurate predictions about the type and timing of environmental events and engage top-down control processes to alter behavior following experiences that violate predictions, then the deprived social environment of institutions may contribute to these deficits. Institutions offer children limited opportunities to learn environmental contingencies in order to facilitate accurate predictions about future events. Moreover, the noncontingent responsiveness of staff might impair children's abilities to predict events after leaving institutional care.

Another possible line of inquiry into brain and behavioral relations is to explore the mechanisms of the association by examining brain functioning in children diagnosed with ADHD. Previous studies of post-institutionalized children have not examined brain functioning in ADHD, so we decided to ask if characteristics of brain functioning might illuminate pathways to signs of inattention and overactivity.

Kate McLaughlin examined whether different frequency bands of the EEG might selectively mediate mental health outcomes.[40] It turns out that they do. We had previously demonstrated significant reductions in alpha-relative power and increases in theta-relative power (both atypical patterns) among children reared in institutions, and we had suggested that these findings were compatible with a delay in cortical maturation, which had been described in previous research as being associated with signs of hyperactivity and impulsivity.[41]

Therefore, we reasoned that low EEG alpha and high theta power, indicating cortical immaturity, might mediate the effect of institutionalization on signs of ADHD. As shown in Figure 11.2, McLaughlin's analyses confirmed that EEG alpha and theta power partially explained the effects of institutionalization on hyperactivity. Results were similar for impulsivity, but there was no mediation on signs of inattention.[42]

Importantly, we also showed that EEG power was unrelated to depression, anxiety, or disruptive behaviors. Thus it appears that the pathway to hyperactivity and impulsivity among institutionalized children may be mediated in part by alterations in brain functioning, which in turn derive from institutional rearing.

Anxiety and EEG Asymmetry

We have known for some time that frontal areas of the cerebral cortex are differentially lateralized for processing positive and negative stimuli and underlie behavioral and expressive responses to emotional information.[43] The left frontal region is activated by pos-

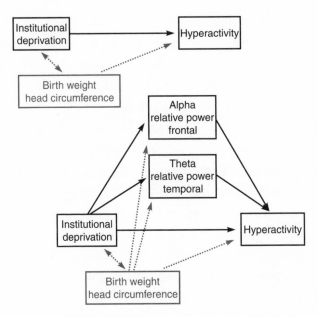

Figure 11.2. EEG alpha and theta power partially explained the effects of institutional rearing on signs of hyperactivity.

itive emotional stimuli and promotes approach behavior, whereas the right frontal region is activated by negative emotional stimuli and underlies withdrawal and avoidance behavior. Previous research had not characterized the development of frontal EEG asymmetry in children nor evaluated whether adverse rearing environments alter developmental trajectories.

Therefore, we examined whether individual differences in relative hemispheric activation of the frontal cortex—as indexed by EEG alpha power—were associated with internalizing disorders at fifty-four months of age.[44] In addition, because of the longitudinal design of BEIP, we explored the trajectories of EEG alpha power over time.

Children with and without a history of institutional rearing demonstrated an initial increase in right relative to left hemisphere frontal EEG activity followed by an increase in left relative to right hemisphere frontal EEG activity. The initial increase in right hemisphere frontal EEG activity occurred for a longer period of time among institutionalized children, however, and the subsequent rebound in left frontal EEG activity was less marked for children exposed to institutionalization. By ninety-six months, children raised in the community exhibited fairly pronounced frontal asymmetry characterized by left-greater-than-right hemisphere EEG activity (indexed by less alpha power in the left frontal lead relative to the right), whereas institutionalized children exhibited a pattern of right-greater-than-left hemispheric activity. In other words, institutional rearing was associated with a prolonged right frontal asymmetry in early childhood, followed by a blunted rebound in left frontal activation in middle childhood. Foster care placement was associated with improved developmental trajectories, but only among children placed before twenty-four months. Importantly, the pattern of frontal EEG asymmetry among the institutionalized group predicted signs of internalizing disorders at fifty-four months.[45]

What are the implications of these findings for our understanding of brain development? Although the specific neural mechanisms underlying frontal asymmetry are not yet known, Richard Davidson has proposed that asymmetry may be associated with positive and negative emotional expression because of inhibition of the amygdala by the left prefrontal cortex.[46] The prefrontal cortex, particularly the right PFC, is known to be involved in regulation of negative emotional responses through the use of cognitive strategies.[47]

Increased activation in the right frontal cortex beginning at about age two may reflect an adaptive developmental change in withdrawal tendencies to novelty as children encounter more environments outside the home and have more interactions with people other than their primary caregivers. A temperamental construct called *effortful control*—including delaying and inhibiting behavior as well as slowing down motor activity—notably increases during this same developmental period and may reflect maturation of the prefrontal cortex.[48] Together with the current findings, these patterns suggest normative increases in behavioral inhibition in two- to three-year-olds that are mediated by developmental changes in the prefrontal cortex.

Placement of children in foster care prior to twenty-four months of age had a positive effect on frontal asymmetry trajectories. The brain circuits underlying hemispheric EEG activation patterns in the frontal cortex seem to have been more responsive to environmental differences in the first twenty-four months, and much less so later. Again, this finding supports the view that institutional rearing profoundly affects some specific aspects of brain development in young children.

Processing Facial Emotions and Internalizing Disorders

As discussed in Chapter 8, the ability to process facial emotion has important implications for effective and appropriate social interactions. In addition, this ability has been shown to be sensitive to the early social environment. Thus children exposed to severe early adversity, including maltreatment and institutional rearing, exhibit disruptions in the neural processing of facial emotion. These disruptions have been observed in both behavioral responses to displays of facial emotion and in event-related potentials (ERPs). Seth Pollak and Pawan Sinha documented greater amplitude of one particular ERP component (the P3b) in response to angry facial expressions in maltreated compared with nonmaltreated children, suggesting heightened allocation of attention to angry faces.[49] Exposure to adverse early-life experiences may have lasting effects on neural processing of facial information. Given that disruptions in the neural processing of facial emotion have been documented in children exposed to adverse rearing environments, these alterations in facial emotion processing may be a neurodevelopmental factor linking adverse environments to the later onset of psychopathology.

We examined whether abnormalities in neural processing of facial emotion, as reflected in ERPs, were associated with subsequent psychopathology and whether these abnormalities in facial emotion processing were a mechanism linking institutionalization to psychopathology.[50] Specifically, we examined the association between several ERP components obtained in a familiar-unfamiliar face task (discussed in Chapter 8) at baseline and mental health outcomes at fifty-four months. We observed that the peak amplitudes of the P100 and P700 components in response to facial stimuli were diminished among institutionalized children compared with community children, and these specific ERP components also were associated with psychiatric symptomatology among institutionally reared children. Reduced P700 peak amplitudes in response to facial stimuli among institutionally reared children partially explained elevations in inattention, hyperactivity, impulsivity, and anxiety at fifty-four months. However, this aspect of neural processing of facial stimuli did not explain the relationship between institutionalization and signs of other disruptive behaviors or depression.

The abnormal neural processing of facial and emotional information partially explained elevations in externalizing and anxiety symptoms among children in the CAUG compared with NIG children. These findings contribute to a growing literature that has begun to implicate specific neurodevelopmental pathways involved in the heightened risk for psychopathology among children reared in institutions.[51]

Conclusion

Previous research had demonstrated strong associations between institutional rearing and psychopathology of various types. In BEIP, we extended this work in several ways.

First, we comprehensively examined the prevalence of psychopathology and impairment using a state-of-the-art psychiatric diagnostic interview. This yielded the most systematic study of psychopathology conducted to date in young children with a history of institutional rearing.

Second, we examined moderators of the intervention, specifically gender, timing of placement, and genetic variability. Unlike most of the results reported in Chapters 7–10, findings for psychopathology revealed powerful differences in outcomes related to gender. Girls benefited from placement in foster care far more than boys, at least with regard to reductions in psychopathology. By contrast, age at placement in foster care had no effect on psychiatric symptoms, disorders, or impairment. We also demonstrated that polymorphic variants in genes known to regulate neurotransmission in key neural circuits underlying depression, attachment, and externalizing behavior moderated intervention effects in predicted ways. Most important, we demonstrated in two separate analyses that the same polymorphisms, potentially part of a larger group of susceptibility alleles, were associated with the most favorable outcomes in the foster care environment and the least favorable outcomes in the continued institutional care environment. This finding is powerful support for a genetic differential susceptibility model that suggests that many common and well-studied polymorphisms are neither good nor bad but rather confer differential ability to adapt to changing environments. In the case of BEIP, this means responsiveness to changes in the caregiving environment.

Third, we examined mediators of the intervention, and we discovered that security of attachment was a powerful mediator of the effects of the intervention on reductions in internalizing disorders. Further, this mediator also explained why girls were more responsive to the intervention than boys—girls were more likely than boys to develop secure attachments. Reasons for this difference need further exploration.

Finally, we demonstrated some novel relations between brain functioning and behavioral outcomes. We showed that EEG alpha and theta power, indexes of cortical maturity, partially mediated the effect of institutional rearing on hyperactivity and impulsivity but not on inattention. We also showed that institutional rearing was associated with a prolonged right frontal asymmetry in early childhood, followed by a blunted rebound in left frontal activation in middle childhood (compared with a more robust left frontal activation among never-institutionalized children). Foster care placement led to a more robust left frontal activation, but only among children placed before twenty-four months. In addition, right frontal EEG asymmetry among the institutionalized group predicted signs of internalizing disorders at fifty-four months.[52] Using ERPs to assess children's reactions to familiar and unfamiliar faces, we demonstrated that amplitudes of specific ERP components were dampened in institutionalized children compared with community children. This is important because these specific ERP components also were associated with selective types of psychiatric symptomatology among institutionally reared children.

Chapter 12

Putting the Pieces Together

We begin with the observation, based on the evidence presented in this book, that exposure to early and profound psychosocial deprivation leads to a derailing of typical development. Why might this be?

The brain's initial architecture and wiring are orchestrated by a series of genetic scripts that begin a few days after conception and continue through the early postnatal period. However, whereas genes provide the framework for the brain-to-be, experiences fill in this outline, leading to the emergence of a mature brain toward the end of adolescence and the beginning of young adulthood.

How exactly does the structure of experience weave its way into the brain? As described in Chapter 1, the neuroscientist William Greenough proposed two ways: experience-expectant and experience-dependent development.[1] Experience-expectant development assumes that infants will encounter environments that are common to all members of the species. Such "shared experience" helps ensure survival and reproductive success, and it also

makes for a more efficient use of genes. Specifically, rather than require our limited genome of approximately 20,000 genes to determine the details of the final form and function of brain and behavior, a more limited set of genes instead constructs a brain that
depends on experience for its full elaboration. Thus for any given
species, there are certain experiences that should be reasonably expected by all members of that species. One example might be how
patterned light information facilitates the development of both
low- and high-level visual abilities (for example, depth perception
and face perception). Another might be how complex auditory information facilitates the development of speech perception. Moreover, moving away from sensory systems to social-emotional development, we know that the physical and emotional availability of
caregivers facilitates the development of attachments.

Inherent in experience-expectant development is the concept
of a sensitive period. For development to proceed normally, elements of the *expectable environment* must be present when the brain
can make best use of this information to assemble itself correctly.
Should the timing be off—for example, a child does not have access
to a particular environment or experience, or the brain or a sensory
organ has been damaged so that it cannot make use of this experience—then development can be derailed, sometimes temporarily,
sometimes permanently. Thus children born with bilateral cataracts
are without patterned light information until these cataracts are
surgically removed. The later in life that surgery occurs, the greater
the child's problems with vision—indeed, past a certain point in
time, a child whose cataracts have not been removed will most certainly have a life-long visual impairment.[2]

The counterpart to experience-expectant development is

experience-dependent development. This form of development is unique to each person and most likely involves the active formation of new synaptic connections throughout the lifespan, on the basis of each individual's interaction with the environment. Examples include learning and memory and the acquisition of a vocabulary, all of which are developmental events that are not bounded by time (that is, we can continue to learn and remember new things all the way through the end of the lifespan, assuming our brains remain healthy).

An overarching truth for both forms of development, according to J. McVicker Hunt, is that "experience cuts both ways."[3] In other words, the nature of the experience itself, coupled with the receptivity of the nervous system at the time the experience occurs, will influence whether the resulting neural change is beneficial or deleterious to the development of the organism. For example, the plastic nature of the infant's brain makes possible the ability to acquire language; that same plasticity, however, will ensure that a child exposed to teratogens in the environment (for example, mercury or lead) will be at risk for atypical development.

We hasten to add that some aspects of development may be both experience-expectant and experience-dependent. As discussed in Chapter 10, attachment may be an example: the *formation* of an attachment may reflect an experience-expectant process but the *quality* of the attachment may reflect an experience-dependent process. A child whose caregivers respond consistently and sensitively during the infancy period is more likely to become a psychologically well functioning adult, capable of forming important, lasting relationships with others. A child who grows up in an institution where "caregiving" largely consists of being fed and changed is far less likely to attain psychological health.

Understanding the Long-Term Effects of Early Psychosocial Deprivation

How might this framework contribute to our understanding of why early, severe deprivation leads to such a profound derailing of development, and why there are such long-term consequences? Simply put, spending one's early life in an institution leads to a violation of the expectable environment. For example, it is not uncommon among infants living in institutions through at least a year of age to spend much of their time lying on their backs in a crib, staring at a white (patternless) ceiling, which does little to stimulate the visual system, such as the movement of the eyes to explore visual patterns. This may explain why we saw so many young children with what looks like strabismus (though we have not formally investigated this observation). Similarly, because of a paucity of caregivers, such infants may have only their basic needs met (for example, being fed or changed) but not their psychological needs. In addition, they may have little exposure to faces or to language. This may lead to delays or disorders in social communication, including the ability to understand spoken language and to "read" faces. In some institutions, adequate nutrition is lacking, eventually leading to profound growth restrictions or, more subtly, to diminished cognitive function owing to specific nutrient deficiencies. Because of the high ratio of children to caregivers, the overall needs of individual children may not be met. Given the weak psychological and neurological foundation upon which subsequent development builds, later developmental goals may be compromised (such as the ability to relate to peers and other adults, or the ability to view the world through the eyes of another). By analogy, if we think of development as a tower of stacking blocks, imagine what happens if the second block is slightly displaced from the first, and the third slightly displaced from the second, and the fourth slightly displaced from the third, and so on. Ultimately the column of blocks begins to lean precipitously, *even if the displacement of each block is only very slight.*

Thus the very foundation of development for children growing up in institutions is often compromised from the beginning, and, over time, further violations in the expectable environment result in further deviations from the initial foundation, ultimately leading the child down a path that is far from typical.

The Bucharest Early Intervention Project demonstrated that profound deprivation occurring early in a child's life may have tragic, long-term consequences for psychological, neurological, and biological development. There is nothing surprising about these findings—deprivation has long been recognized to be harmful. Still, BEIP represents the most comprehensive, systematic, and detailed study ever conducted of the brain and behavioral development of children being raised in institutions, and it has provided the most detailed documentation to date of deficiencies. With very few exceptions, we demonstrated compromises in virtually every aspect of development among institutionalized children. From the level of molecular structures to the level of complex social interactions, from brain structure to brain function, these children are unquestionably disadvantaged.

Gauging Results of the Intervention

But how much of what we see in these children results from being raised in institutions and how much from characteristics that predate their institutional rearing, such as genetic dispositions or prenatal experiences? The only way to know was to compare the development of children who share these same risk factors (that is, what may have led them to be institutionalized to begin with) but who had a different postnatal experience. This was a major aim of BEIP—to determine if removing infants and toddlers from institutions and placing them in foster care would augment their progress in various domains of development. By randomly assigning children to foster care, we were able to determine that the differences between the FCG and the CAUG were due to the foster placement. Nevertheless, it is important to remember that the children were assigned to foster care not at birth but rather after a period of between seven and thirty-one months of institutional rearing. In fact, the mean age at placement was twenty-two months. Thus any differences we demonstrated between the children in foster care and the children in institutions are limited by the fact that the foster care group had already experienced at least some institutional rearing.

Another factor that may restrict the magnitude of the effects of the intervention is the intent to treat approach, in which we analyzed data on the basis of treatment *assignment* rather than treatment received. Here, it is worth recalling that at the eight-year follow-up, only fifteen of the sixty-eight children in the care-as-usual group remained in institutions. This means that differences between the FCG and the CAUG are conservatively reported, since the majority of the CAUG were also living in families. In the two domains in which we examined the children who had remained in institutions throughout the study—IQ and reactive attachment disorder—the continuously institutionalized group was even more compromised.[4]

Given our findings, BEIP demonstrated that placing institutionalized children into families ameliorates many, though not all, of the negative outcomes of early institutionalization and likely under-states the potential power of family placement. This is especially important in our subsequent discussion of policy implications.

Of course, not all domains were compromised in institutional-ized children (for example, ERP response to stranger versus famil-iar adult) and not all domains of development responded to the in-tervention (for example, executive functions). Also, in almost no domains was the functioning of the foster care group comparable to the functioning of the NIG, so the recovery was incomplete, at least through age eight. These qualifications about the findings must also be considered in any conceptual framework.

Sir Michael Rutter, who has long been interested in the effects of early experience on later development, has proposed that two variants of *developmental programming* are responsible for the effects of early institutionalization on later development.[5] Building on Greenough's work, Rutter argued that when a child's experiences fall far outside the expectable range of environments, as occurs in institutions, development can be derailed—what he refers to as *experience-expectant programming*. In contrast, *experience-adaptive pro-gramming* reflects a very different phenomenon: the child's behav-ior in the institution, however abnormal outside the institution, is adaptive within the institutional environment. Thus the indiscrimi-nate behavior that we and many others have characterized among institutionalized children may be considered adaptive in a world in which adults pay little attention to children, and it falls to children to solicit as much attention as possible from adults—even if that means hugging a complete stranger.

Institutionalization, as we have argued, represents a form of ne-glect, albeit a very extreme form. Nevertheless, the experiences of children raised in institutions are most likely quantitatively rather than qualitatively different from what is commonly observed among children in many different child-protection systems: they lack supervision and structure, necessary social interaction and emotion regulation, cognitive and language stimulation, access to adequate nutrition, and so on. Neglect can provide a framework for understanding how institutional care can compromise the developing brain.

If we reflect back to Chapter 1, we can see how devastating profound neglect is for brain development. If genes specify only the rudiments of the brain-to-be, on the assumption that experience will bear responsibility for the major proportion of development, imagine what happens when such experiences fail to occur. As much as the brain needs oxygen, it also needs experience, and if it is starved of experience, it is left to its own devices to continue its developmental journey—a journey that has little structure, little input, and an inadequate road map. Neurons and synapses that await confirmation from the environment (the "use it or lose it" phenomenon) meet an uncertain fate, perhaps leading to over-pruning, in some cases, and a lack of pruning in others. The end result could be a brain with too few neurons and connections between neurons. Indeed, the EEG and MRI data support this hypothesis: recall that the institutionalized children had smaller heads, under-powered brains, and reduced total cortical volume (including reductions in both gray and white matter) than never-institutionalized children.[6]

Of course, in the case of the BEIP, it is not only the lack of critical experiences occurring at specific time points that contributes to the outcomes we observe; it is also that some of the experiences themselves may be detrimental to development. For example, the institutionalized children were fed by a caregiver, which is an expectation of our species' young. But *how* they were fed—having food shoveled into their mouths while they faced away from the caregiver—was not an experience conducive to developing a warm and trusting relationship.

Simply put, children living in institutions lacked input, and the lack of input led to maladaptive brain development. In addition, in some cases the input they received was in violation of the expectable environment—thus it is not just inadequate but also damaging. The extent to which the harm that results from lack of input and faulty input can be compensated for or corrected will be influenced in part by the degree of deprivation and the duration of deprivation. As many studies have now shown, the sooner a child leaves an institution, the better.

Care in Families

The BEIP revealed not only that early deprivation negatively affects development, but also that children placed in families after being reared in an institution do better than those left behind in the institution. Why might this be? Very simply, families (at least, the families in our intervention) provided for the child's expectable needs. Children were talked to, played with, held when needed, engaged with, encouraged to master their world, respected, and loved. These are just some of the things our species has come to expect, and though families may vary in composition, size, and even the ability to provide consistent, high-quality care, the point is that *most* are able to provide these basic needs to varying degrees. Our ratings of observed caregiving behavior thus revealed a substantial difference between families and institutions. Naturally, some children who grow up in families still experience developmental or psychological challenges, but even these individuals are still far more likely than children raised in institutions to lead meaningful lives and contribute to society.

The investment that loving adults can make in an infant is missing in institutional care for a number of reasons. First, caregivers in institutions are responsible for many children and have many tasks to handle. As a result, they may not appreciate the needs of each individual child as important. Second, many caregivers have little formal education (the educational requirement to become a caregiver in Bucharest was an eighth grade education) and little training in child development. Thus they may not know precisely how to invest in a child's needs. Third, they are paid low wages, meaning that they may feel undervalued, and otherwise qualified individuals may not consider doing this work. This is not to say that formal education, more adequate compensation, or even training in child development are essential to fostering healthy development. But *in the context of taking care of vulnerable children,* we know that training may contribute to improved outcomes.[7] Finally, because of the widespread beliefs associating institutional rearing with defective children, caregivers may invest little in the children for whom they are responsible, as they are likely to believe that the child's development was determined by the circumstances that led him to be placed in an institution to begin with.

Overall, then, child rearing in families differs in fundamental ways from child rearing in institutions. Like institutional care, family care varies from one family to another, but assuming that we are dealing with typical families, they will all provide for a child's "expectable needs" to varying degrees. Viewed this way, being raised in a family is categorically different from being raised in an institution.

Timing Matters

Detecting effects related to the timing of placement in foster care is quite consistent with the literature in neuroscience on sensitive periods—not all circuits are governed by sensitive periods, but of those that are, the precise timing may vary from domain to domain. Also, for the complex behaviors in which we are interested—attachment, language, and psychopathology, for example—many circuits are probably involved in the expression of the behavior, and they may have many different sensitive periods.[8] In our work, for example, the sensitive period governing language development differed from that underlying intellectual development. This means that the neural circuits subserving language most likely require input at a different point in development than those subserving intellectual or emotional development.

Because children were placed in foster care at a range of ages—seven to thirty-three months—we were able to assess the effects of timing of placement on outcomes at each age of assessment and follow-up. As shown in Table 12.1, we found that in some but not all domains of development, institutionalized children who were placed in families before certain ages did better than those who were placed later.

Early in development we found evidence for sensitive periods across a wide range of outcomes. At forty-two months of age, IQ, expressive and receptive language, stereotypies and security, and organization of attachment all showed these effects. The question, of course, is could these effects be sustained over the course of development?

Why did some domains not show a sensitive period, such as the reduction in signs of internalizing disorders (that is, reductions in

Table 12.1 Timing effects for foster care placement evident in assessments at 42 months, 54 months, and 8 years

	42 months		54 months		8 years	
	Intervention effects	Timing effects	Intervention effects	Timing effects	Intervention effects	Timing effects
EEG power	no	no	n/a	n/a	yes	yes
IQ	yes	yes	yes	yes	yes	no
Language	yes	yes	n/a	n/a	yes	yes
Stereotypies	yes	no	yes	yes	n/a	n/a
Social skills	n/a	n/a	n/a	n/a	yes	yes
Selective attachment	yes	yes	n/a	n/a	n/a	n/a
Height and weight	yes	yes	n/a	n/a	n/a	n/a

anxiety and depression)? First, it may simply be that these functions are experience-dependent and thus will improve regardless of when the environment changes. No matter how old a child is when placed in a family, then, that child will experience a reduction in anxiety and depression (though even here we must be careful, because no child was *older* than thirty-three months when placed in foster care, so we are unable to say whether children placed into families past this age will benefit as much as those placed at or below this age). If so, this suggests that these functions are experience-dependent, not experience-expectant. Second, though children's internalizing symptoms were reduced by placement in a family, this is not to say that we might not have witnessed an even greater improvement had the children been placed far younger than we were able to place them (say, within the first six months). In other words, our experimental design may have lacked the sensitivity to detect the full range of sensitive periods. A third possibility is that improvement in these domains was secondary to improvement in other domains. For example, perhaps improvements in cognitive function led to a reduction in anxiety and depression—thus the changes we saw in internalizing symptoms were secondary to placement in foster care, not primary. After all, improved cognitive ability plausibly could bring with it improvements in the ability to regulate one's internal mental states.

Unfortunately, a shortcoming of our experimental design is that we had few children who were very young at the time of placement in foster care. As a result, we are unable to offer very precise timing information; we can, however, say with great certainty that for a number of domains (for example, language, some aspects of cognition, and security of attachment), the sooner children are placed in families the better their chances for subsequent healthy adaptation. This also is consistent with the work of other researchers studying children who have been raised in depriving institutions.[9]

Timing Matters but Perhaps Not at All Ages

The concept of a sensitive period brings with it the assumption that if a certain experience does not occur during a particular window of time, the circuit and its corresponding behavior will be altered permanently. If such change occurs early in life one possibility is that it forms the foundation for more complex related behaviors that emerge afterward. In such a case, subsequent complex behaviors will continue to display the same sensitive period effect as the simpler earlier behavior. A number of alternative pathways are possible, however. One is that while a sensitive period may exist early in life, evidence of its effect may be detectable only later in life. This is known as a *sleeper effect,* an effect that does not emerge until later in development, though it is based on early experience at a particular sensitive period.[10] Another possible trajectory is that environmental circumstances may create the conditions under which circuits formed early in life during a sensitive period no longer work as efficiently as they once did. In our own data we have evidence for both of these alternative trajectories.

Evidence of a sleeper effect emerged in EEG and brain-imaging results. Recall that at baseline the institutionalized children showed significant reductions in typical EEG power compared with the community children. Although we continued to measure EEG power at thirty and forty-two months after the intervention had begun, it was not until age eight that both an intervention effect and a sensitive period were evident. That is, children placed before twenty-four months of age had EEG power comparable to that of community children.[11]

An example of the second trajectory occurred in the IQ results at age eight. At forty-two and fifty-four months we provide clear evidence for sensitive periods in IQ, with children placed younger than twenty-four months having higher IQs than those placed after twenty-four months. However, at eight years, though we continued to observe an intervention effect, we no longer observed a timing effect. This could be due to one of several reasons. First, it may be that cumulative life experience gradually came to play a greater role in the skills necessary for a complex set of cognitive processes assessed with IQ, overriding the effects of early experience. We found that among children still living in BEIP foster homes, earlier placement was related to higher IQ. If we assume that the various cognitive abilities that constitute IQ are largely experience-dependent (that is, influenced by the learning that takes place across the lifespan), then perhaps we shouldn't be surprised that as children grow older, their current life circumstances play a greater role than their early life experiences.

Not All Domains Responded to Intervention

A few domains of development did not show intervention effects. For example, we had little success improving executive functions or reducing the prevalence of externalizing behaviors such as ADHD. Why might this be? We have several hypotheses. First, it may be that we placed children too late, and, had we placed them much earlier, we might have been more successful in getting them back on course. In the English and Romanian Adoptees Study, children adopted from Romanian institutions before the age of six months seemed almost indistinguishable from domestically adopted children in the UK.[12] Second, it could be that the neural circuits that support certain behaviors (for example, inhibitory control) were compromised prenatally or even in the very early postnatal period, and as a result, no matter how young the child was when placed in foster care, we would not have been successful in righting the course of development. Third, it may be that spending even a few months in an institution is enough to disrupt the neural circuitry that subserves these behaviors, and that only intensive and intentional therapeutic efforts as opposed to generically more positive caregiving, as in our study, would be able to remediate deviant development. For example, at fifty-four months, externalizing disorders were prevalent and largely unresponsive to the intervention. Specific evidenced-based treatments for externalizing behaviors, such as Parent Child Interaction Therapy, may be necessary to right these children's developmental trajectories.[13] If these are experience-dependent circuits, we must make a concerted effort to figure out precisely which experiences drive development.

Of course, a fourth, more optimistic hypothesis is that the early sensitive periods gradually are overwritten/modified/shaped/remodeled by subsequent life experience. Such experience might take the form of compensatory processes, such as those that operate in adults following brain injuries (such as strokes or trauma). We might posit here, then, that the principles of neural plasticity come into play, allowing the brain to remodel itself. Whether such compensatory processes can override the harm done during early development is a subject of great importance, but it will require following our sample as they make the transition to adulthood.

Normalizing Development?

There were few domains in which the institutionalized children ever looked "normal." For example, if children were placed in foster care before two years, they experienced an increase in IQ, but the mean IQ scores were in the low-average range (among children placed in foster care between seven and eighteen months, the mean IQ at fifty-four months was 85). Similarly, though foster care greatly reduced the prevalence of anxiety and depression, it did not reduce the levels to those of the never-institutionalized group. Why was the intervention incompletely effective?

We have considered several possibilities. First, perhaps we placed the children in foster care too late; had we placed them after only a few months of age, maybe they would have looked indistinguishable from the NIG. Second, there may have been subtle biological differences between children abandoned to institutions compared with those who were kept by their families, such as whether the mother had prenatal care, or whether the infants experienced pre- or perinatal complications. For example, our record review and pediatric exam could have shown lead or other toxin exposure, iron deficiency anemia, prenatal substance exposure, and so on, and these biological "brakes" might limit children's recovery. Indeed, we do have evidence from our work that children who have more signs of biological insult (greater prematurity, lower birth weight, greater prevalence of stereotypies, smaller head circumference) may be more impaired than those without such signs. Unfortunately, our sample size is too small to examine this issue in sufficient depth to answer the question definitively.

Third, children may not have fully recovered because they carry genetic polymorphisms (see Chapter 11) that in conditions of deprivation predispose them to developmental limitations. For exam-ple, children born with two versions of the short allele of the 5HTT (the serotonin transporter gene) may be more vulnerable to developing anxiety if they are brought up in an institution. These polymorphisms may affect how completely children recover once placed in a family.

Fourth, perhaps we should not expect full recovery in all do-mains, for the simple reason that the forces that drive development do so selectively and operate differently in different domains. A case in point is the contrast between IQ, which never fully recovered, and social-emotional development, which at least among children placed in foster care before two years old did largely recover.

Finally, we must consider that though BEIP foster care was of reasonably high quality, it was perhaps insufficiently therapeutic to remediate some developmental delays and deviances. Given our observations made at baseline, many of the children were seriously delayed or handicapped in multiple domains, and as a result, some (for example, those with signs of ADHD or cognitive impairments) may well have benefited from more targeted interventions.

Policy Implications

We begin by acknowledging the uneasy relationship between sci-ence and policy. Scientists are trained to be skeptical and quick to qualify and note limitations about new findings, even their own. The oft-stated conclusion of research papers in behavioral science is, "More research is needed." This is less about perpetuating the research enterprise and more about acknowledging that no single study can answer truly important questions definitively—at least not to the satisfaction of researchers. Policy-makers, by contrast, demand conclusions—black or white distinctions. Either we invest our resources in a particular effort, or we do not. And since resources are limited, decisions often have to be made definitively. Further, a number of factors other than science affect policy decisions, including political ideologies, economics, cultural values, human rights, and other related perspectives. We believe that science can inform policy by providing more objective data about potential areas of investment and especially about preferred forms of intervention.

In the remaining sections of this chapter, we consider the implications of BEIP and related studies for policy-makers, emphasizing three major points: 1. family rearing is clearly preferable to institutional rearing; 2. the younger a child is placed in a family the more likely the child is to have a better outcome; and 3. the quality of the alternative placement matters considerably.

Institutional Rearing

Estimates are that 2,000,000 to 8,000,000 children worldwide are raised in institutions; indeed, we have noted that this figure likely underestimates the real number, given the challenges UNICEF and other organizations have had in documenting the existence of institutions. Some governments may under-report the number of institutions or classify them in such a way as to disguise the fact that they are, in fact, institutions. Others may simply lack the infrastructure or the commitment to collect meaningful data.[14] In Central and South America, Asia, the Middle East, and Africa, and even parts of Europe, institutional placement of orphaned, maltreated, and abandoned young children is still common. Ten case control studies published in English comparing children in foster care with those in institutions in four different countries all have consistently shown that children in foster care develop more favorably.[15] On the one hand, this is a relatively small database, including some limited and mostly descriptive studies, but on the other hand, the consistency of results across settings and decades is notable.[16] The caveat, of course, is that none of these studies included randomization, making it possible that more handicapped children are placed in institutions than in foster care. This means that better outcomes in the children placed in foster care may derive from preplacement factors.

We heard one form of this argument in Romania. It went something like this: Of course the institutionalized children are delayed, because problems that they had prior to institutionalization compromised their development. Further, because they are damaged, there is little reason to spend limited resources on them. This position, as we discussed in Chapter 3, has its roots in defectology, the Soviet idea that intrinsic abnormalities would not be amenable to intervention.

BEIP is a strong refutation of the argument that preplacement factors accounted for all the differences in children in the foster care and the care-as-usual groups. Because of randomization, the possibility of selection bias in rearing environment was eliminated. Therefore, BEIP has provided the strongest evidence to date that family placement is more advantageous than institutional rearing.

Policies supporting the advantages of family rearing for young children should be developed where they are lacking and implemented fully if they are not. But policy rarely derives directly from scientific findings. As noted, institutional rearing is quite common in most parts of the world. A detailed analysis of the reasons this is the case is beyond the scope of our study, but we briefly consider several relevant issues.

In the United States, the orphanage movement intensified at the same time prisons and asylums for the mentally ill began to be used

widely. In the latter part of the nineteenth century, when industrialization, urbanization, and economic downturn contributed
to large numbers of homeless children and adults, all of these institutions were considered strikingly innovative societal interventions.[17]

Disparate beliefs supported these new models. For some, weak-
nesses in the character of abandoned or wayward children required
remedial efforts. For young children this meant being raised in stable but regimented environments in which they could be properly
educated and trained under the control of authorities. For others,
keeping wayward children who came from troubled parents out
of sight (and mind) reduced any threats they might pose to the
larger society.[18] Thus forces of remediation and of pragmatism converged in supporting the institutionalization of young children in
the United States in the late nineteenth century. Once established,
of course, orphanages were perpetuated by economic forces, as
money was allocated on the basis of need and secondary benefits
accrued to those involved in such systems, since they provided jobs
and other benefits to communities.

In Romania, both remedial and pragmatic forces also seemed to
have perpetuated institutional care for young children. These forces
were still palpable as we began our work there at the end of the
twentieth century. Ceauşescu's authorities had promulgated the notion that state-trained orphanage staff could be more effective raising children than poorly educated and troubled parents. We heard
this remedial argument from a number of individuals working
within institutions. The Romanian version of the pragmatist position, by contrast, may have actually predated the Communist era,
but it was still present afterward. Lavinia Mitroi noted that in the
Eastern Orthodox tradition, children who had physical deformities,
handicaps, and disabilities were sometimes considered a physical

manifestation of some sin committed by either parents or the family and represented physical evidence of God's punishment.[19] The stigma of having a handicapped child may well have been another contributor to the placement of some children into institutions.

Getting abandoned or handicapped children out of sight and out of mind, whatever it derived from, was largely successful in Ceaușescu's Romania. Repeatedly, we were struck by how many people who had worked in Romania in the Communist era now say that they were unaware of the numbers or sometimes even the existence of large institutions for young children. Keeping abandoned children in Romania in institutions kept them out of sight, meaning that the painful questions about what to do about such children did not have to be confronted, particularly at a time when virtually everyone was struggling economically.

No doubt there are other reasons that institutions for raising young children persist in many parts of the world beyond Romania. Agreeing to care for children who are not biologically related may well require a benevolence in adults who could potentially serve as foster or adoptive parents—and the necessary sense of well-being that could create benevolence may be scarce when adults feel uncertain about their own families. That is, in countries with pervasive disadvantage and limited access to resources, adults may not feel economically secure enough about themselves and their families to accept the additional burden of caring for other people's children. This could explain, at least in part, why low-resource countries seem to rely more on institutional care for abandoned children than do high-resource countries.

Timing of Family Placements

For many domains of development, earlier placement of young children in families confers additional advantages. This is clear from many of the findings we have summarized in Chapters 7–11. But it is also clear from adoption studies conducted from the 1970s to the 1990s.[20] Although we and others have highlighted specific ages after which developmental compromise is more pronounced, a comprehensive review of relevant studies indicated that data on this question most clearly support the "earlier the better" rule for removing children from institutions rather than at any specific age.[21] That is, the older the child the greater the risk for impairment across multiple domains of development.

We know that it is easier and more efficient to "build" a brain correctly the first time than it is to rebuild it following injury or adversity. Thus, waiting to intervene until after a brain is largely developed will require more effort (and likely more money), and successful remediation is far less likely. Our findings accord well with longitudinal intervention studies in disadvantaged children, neurobiological studies in animals, and economic analyses of early intervention. Eric Knudsen and colleagues selectively reviewed these disparate literatures and concluded that early experiences have a uniquely powerful influence on the development of brain architecture and skill development.[22] They asserted that the capacity for change in human skill development and neural circuitry is highest earlier in life and decreases over time. Therefore, the convergence of findings they cited strongly supports investing in enhancing the environments of disadvantaged children during the early childhood years. These conclusions are fully supported by our work in BEIP.

Alternatives to Institutional Rearing

Another implication of results from BEIP is that attention must be paid to the quality of the alternative placement for the child. As we have noted, the more the setting resembles a family, the more the child's development will be facilitated. Short of placement with biological relatives, the most family-like setting for institutionalized children is foster care. Although investing in high-quality foster care seems obvious, it is far more challenging to develop a *high-quality* foster care system for young children than to merely develop a foster care system.

We have data from outcomes including both psychopathology and IQ that illuminate the importance of high-quality foster care versus "average" foster care. We showed that children in BEIP foster care placements at age fifty-four months had substantially lower levels of psychopathology than children in government-sponsored foster care. Further, we also have reported significantly higher IQs at age eight for children in BEIP foster care compared with children in government-sponsored foster care.[23] This is a direct comparison of a carefully designed foster care intervention that was overseen by well-trained social workers with caseloads of 18–20 families each. Many of the Romanian government social workers told us in the mid-2000s that they had 80–100 families to supervise. This, of course, is a formula for failure. Furthermore, the focus of facilitating healthy attachment relationships between young children and their caregivers was a central component of BEIP foster care, but it is not even considered in typical government-sponsored foster care.

This is not meant to be an indictment of Romanian governmental foster care. The child-protection challenges that Romania faced in 1990 were staggering and unprecedented. More than twenty years later, progress has been considerable by many metrics. At the same time, Romania is still far from a developmentally informed child-centered system of caring for abandoned children that we and others have advocated as the ideal to strive for, and that clearly does not exist as yet in the United States.[24]

The Question of International Adoptions

As described in Chapter 3, Romania became embroiled in a contentious dispute between the European Union, which opposed international adoption, and the United States, which supported international adoption. This conflict continues, as Romania temporarily banned international adoptions in 2001, permanently banned adoptions in 2005, and then in 2012 passed legislation that allowed for international adoptions of Romanian children by fourth-degree relatives, by spouses of the child's natural parent, and by Romanian citizens who habitually reside abroad.

Much of the opposition to international adoption stemmed from rampant corruption in the process in Romania in the first decade after Ceauşescu. In addition to child-protection authorities' arranging adoptions for bribes, there were allegations of poor women being paid to conceive, of adoption of children whose parents retained their legal rights, and other horrific scenarios. Rumors spread of children being adopted for purposes of organ trafficking, prostitution, and pedophilia. Shihning Chou and Kevin Browne have asserted that children's rights are frequently violated in international adoptions, and they cite the Hague Convention (1993), noting that international adoptions should only be considered an option of last resort.[25] They object to the "marketing" and profit aspects of international adoption, and they use inflammatory language to describe "importing and exporting" countries. There are also some legitimate issues not raised by Chou and Browne, especially the danger that internationally adopted children will lose important aspects of their culture, language, and identity.

The U.S. government, by contrast, has been a vocal opponent of Romania's efforts to eliminate international adoption. It argued that there was no justifying the elimination of international adoption as a child-protection option given that tens of thousands of children were being raised in institutions and thousands more continued to be abandoned each year in Romania. Furthermore, adoption achieves the laudable goal of permanence for children, a strong emphasis in U.S. legislation during the last fifteen years (Public Law 105–89). Foster care was always intended to be a short-term intervention, especially for young children. Therefore, also arguing from the position of children's rights, the United States has come to a conclusion very different from that of the European Union about the desirability of international adoptions. Emphasizing permanence for young children, the United States considers international adoption not an option of last resort but an option that is far better than long-term foster care or institutional rearing. And, in fact, the Hague Convention favors international adoption over institutional rearing.

BEIP's major contributions to this debate are threefold. First, we demonstrated clearly that family care is preferable to institutional care. Second, we demonstrated that quality of caregiving in any setting matters considerably for the child's developmental trajectory. Because higher-quality caregiving occurs in families, policies supporting these placements should be prioritized over any type of institutional rearing. Finally, the timing effects we have observed in BEIP should serve as a caution to the child-protection authorities that time is of the essence in moving children from adverse environments into healthy environments.

Is There Any Role for Institutions in Child Protection?

In some places in the world, of course, foster care is nonexistent and unlikely to be implemented in the foreseeable future. AIDS orphans of many sub-Saharan African countries do not have access to foster care, for example. The question is whether in such settings there is any constructive role for institutions. We believe that there is sufficient evidence to suggest that the kind of large institutions that were widespread in Romania in the last part of the twentieth century have no role in an effective system of child protection. Evidence to date suggests that the more the settings in which children are raised resemble families, the greater the chances for children to develop healthy brains and behavior. Furthermore, a clear lesson from Romania is that once a country invests in institutions, economic pressures build that self-perpetuate the system. Small villages in Romania depended on institutions for jobs, and local and even national governmental agencies and departments received resources on the basis of the numbers of children cared for in institutions. As described in Chapter 3, these economic forces were significant barriers to reform in Romania.

Unquestionably, all efforts should be made to place abandoned or orphaned children with families, ideally in permanent placements. In places where permanent care cannot be obtained, high-quality foster care should be considered. Lacking that, small group settings emphasizing stability of caregivers and restricting the numbers of caregivers involved with each child should be considered. Mixed age groups of children will also make these settings more family-like. Deviations from these approaches will likely increase risks for young children sufficiently that we do not believe that they warrant investment, particularly because of the tendency of institutional interventions to become self-perpetuating.

New Directions in Child Protection

The story we have told has applications well beyond Romania or even Eastern Europe. "Deprived" is a term that may be used not only to describe specific settings for children but also to characterize the insufficient systems of child protection that exist throughout the world. Although we note a continuum of efforts to intervene effectively with orphaned, abandoned, or maltreated children —ranging from children living on their own on the streets, to large and impersonal institutional care, to smaller group settings that may be "home-like," to foster care placement, to augmented foster care —problems abound in virtually all these efforts. Our collective experience is that few if any places in the world can feel satisfied with their child-protection efforts. Limited funds, inadequately trained staff, lack of developmental orientation, failure to consider children's best interests—these are themes that resonate from Bucharest to Boston to Beijing.

Recently, concrete steps have been taken in Romania, in the European Union, and in the United States that have made use of the findings of BEIP, as well as other data, that we believe suggest a shift toward more coordinated and integrated child-centered practices in child protection.

Romanian Government Efforts

The past fifteen years have witnessed remarkable changes in child-protection policies in Romania. As Cristian Tabacaru noted, there were essentially no foster homes in Romania before 1997, and now there are tens of thousands.[26] There have been large strides in de-institutionalization, but smaller strides in preventing abandonment. In 2005, the Romanian government passed Law 272/2004, which made it illegal for children to be placed in institutions in the first two years of life, unless they were handicapped. Although we are unsure of the precise role (if any) that BEIP played in contributing to these policy initiatives, they did coincide with our presentation of our initial results to the government and publication of the first scientific studies on BEIP.

Gabriella Koman, who directed the National Authority for

Child Protection in Romania from 2000 to 2004, said that she and her staff were aware of and affected by BEIP's early results, but she also noted that many factors contributed to reforms.[27] Bogdan Simion, currently executive director of SERA Romania and president of the Federation of Non-Governmental Organizations for Children, also believes that BEIP has been one important source of data in the past decade of ongoing reforms in Romania.[28] Others have made similar assertions.[29] Results of BEIP, of course, are still emerging, and we hope to provide data that will be increasingly useful to policy-makers in Romania in coming years as well.

U.S. Government Efforts

Following the 2010 earthquake in Haiti, the United States government responded with aid but soon realized that its efforts were uncoordinated, often duplicative, and sometimes even counterproductive. Other, similar experiences of lack of focus, coordination, and integration of aid efforts led to a comprehensive review by all of the U.S. government departments and agencies working on behalf of the world's most vulnerable children. Offices within the Departments of Agriculture, Defense, Health and Human Services, Labor, State, the Peace Corps, and the United States Aid for International Development (USAID) were included. The review indicated that the U.S. government provided more than $2,000,000,000 in fiscal year 2009 for about 2,000 projects to assist vulnerable children and their families in more than 100 countries.[30] The review concluded that interventions needed to focus more intentionally on building sustainable child-protection systems that effectively addressed the needs of vulnerable children outside of family care.[31]

One of the results of the review was that the government organized an "Evidence Summit" designed to review evidence on "Protecting Children Outside of Family Care" as a U.S. government interdepartmental initiative. The summit was held in December 2011 and included researchers, international child-protection experts, and U.S. government personnel who comprehensively evaluated the available evidence base for interventions and policies for children outside of family care in low- and middle-income countries. A major emphasis was to determine what additional evidence was needed in order to inform U.S. international aid efforts, but BEIP was featured prominently as an example of the kinds of evidence critical to informing effective policy development.[32] The U.S. government recently released its strategy to promote informed "best practices" for orphans and vulnerable children outside of family care.[33]

European Union Efforts

In June 2012, the European Union Parliament organized an Exhibition and a Roundtable on Children in Institutions. Chaired by Malread McGuinness, MEP from Ireland, the roundtable included presentations on children's rights, BEIP and the developmental effects of institutional rearing, the economics of institutions, and practical implication of reforms.

NGOs such as ARK (Absolute Return for Kids) and Eurochild co-sponsored the event, which intended to highlight the EU's interest in supporting the elimination of institutions as a method of child protection within its borders. This is a notable commitment because research conducted in this decade has shown that institutional care for young abandoned children was common in many Western European counties as well as in Eastern European countries.[34] One consensus of the meeting was that "structural funds" provided to less advantaged EU countries should be used to assist these countries in developing alternatives to institutional care.

Policy Conclusions

From the First White House Conference on Children in 1909 to the United Nations Convention on Rights of the Child in 1989 and the Hague Convention in 1993, the importance of placing children in families has been asserted as a guiding principle. These assertions were informed primarily by values. BEIP and related research have created a solid evidence base for this position.

Careful science has demonstrated conclusively that it is easier and more efficient to "build" brains correctly from the outset than it is to rebuild following injury or adversity. Waiting to intervene until after brains are more developed means that more effort and more resources will be required and chances for successful outcomes will be reduced. Thus policies supporting early intervention efforts for young children raised in deprivation are clearly validated by BEIP and similar research.

Furthermore, with its emphasis on experience, BEIP has implications for the nearly half a million children in the United States who are in foster care. Outcomes in BEIP depend on early experiences, but subsequent experiences also matter. Too often, we forget that foster care is an intervention for children who have experienced maltreatment. U.S. child-protection efforts are geared toward family reunification, as they should be, but there is far too little attention paid to the children's experiences while in foster care. Our research suggests that children's outcomes are likely to be critically related to the experiences they have in foster care. BEIP has revealed the many ways brain and behavioral development are tied to experience, and we should draw from this work, and the work of other scientists, to create a child-protection system that is more rooted in what we know about the "brain-experience" interface than it is about politics or what feels right.

Deprivation, Brain Development, and the Struggle for Recovery

This book has been about a project that for more than a decade has followed children who shared experiences of serious and species-atypical adversity. Following an external manipulation that placed half of the children into families, two groups resulted, and they began to follow different developmental trajectories. The lessons of BEIP—at least through the first eight years—are several. First, the development of children raised in institutions is profoundly compromised. Some of the compromise probably predated the institutional experience—genetics and adverse pre- or perinatal experiences—but from our work and the work of others, we know that many of the abnormalities are due to institutional rearing. Second, within the group of children placed in foster care, earlier placement into families was meaningfully associated with better outcomes for some (although not all) measures. This finding stands out as among the most significant from this work, and it has powerful implications for children worldwide who are exposed to adversity early in life. Third, placement of children in families ameliorates some but not all of the developmental abnormalities. As we have emphasized, we have been conservative in our approaches to analyzing and reporting treatment effects, as is appropriate from a scientific perspective, though we may well have underestimated intervention effects. Finally, the shadow of the children's earliest experiences remains evident. That is, few of the original sample achieved recovery that was comparable to their never–institutionalized peers, at least through eight years, but given the magnitude of the deficits, we do not expect this to change even with the kind of neural reorganization that occurs in adolescence.

Our story was set in Romania, which has particular historical and contextual characteristics that we have highlighted. Nevertheless, the similarities of the Romanian children we studied to children in other countries who have experienced serious deprivation are more important than their differences. We look forward to translation of evidence from clinical, developmental, and brain sciences to construct systems of child protection that provide comprehensive, sustained, developmentally informed, and culturally appropriate care. Only then can we meet the challenges posed by serious deprivation and facilitate children's struggle to recover.

References

Abela, J. R. Z., B. L. Hankin, E. A. P. Haigh, T. Vinokuroff, L. Trayhern, and P. Adams. 2005. Interpersonal vulnerability to depression in high-risk children: The role of insecure attachment and reassurance seeking. *Journal of Child and Adolescent Psychology* 34: 182–192.

Adolphs, R., and M. L. Spezio. 2006. Role of the amygdala in processing visual social stimuli. *Progress in Brain Research* 156: 363–378.

Ahmad A., J. Qahar, A. Siddiq, A. Majeed, J. Rasheed, F. Jabar, and A. L. von Knorring. 2005. A 2-year follow-up of orphans' competence, socio-emotional problems and post-traumatic stress symptoms in traditional foster care and orphanages in Iraqi Kurdistan. *Child Care Health and Development* 31: 203–215.

Ainsworth, M. D., M. C. Blehar, E. Waters, and S. Wall. 1978. *Patterns of attachment: A psychological study of the Strange Situation.* Hillsdale, NJ: Erlbaum.

Albers, L., D. E. Johnson, M. Hostetter, S. Iverson, M. Georgieff, and L. Miller. 1997. Health of children adopted from the former Soviet Union and Eastern Europe: Comparison with pre-adoptive medical records. *Journal of the American Medical Association* 278: 922–924.

Allen, J. P., C. Moore, G. Kupermine, and K. Bell. 1998. Attachment and adolescent psychosocial functioning. *Child Development* 69: 1406–1419.

Almas, A. N., K. A. Degnan, A. Radulescu, C. A. Nelson, C. H. Zeanah, and N. A. Fox. 2012. The effects of early intervention and the moderating effects of brain activity on institutionalized children's social skills at age 8. *Proceedings of the National Academy of Sciences* 109: 17228–17231.

Almas, A. N., K. A. Degnan, O. L. Walker, A. Radalescu, C. A. Nelson, C. H. Zeanah, and N. A. Fox. In preparation. The effects of early institutionalization and foster care intervention on children's social behaviors at age 8.

Ames, E. W., and M. Carter. 1992. Development of Romanian orphanage children adopted to Canada. *Canadian Psychology* 99 (6): 444–453.

Bachman, R. D., ed. 1991. *Romania: A country study.* Washington, DC: Library of Congress, Federal Research Division. Available at http://lcweb2.loc.gov/frd/cs/rotoc.html.

Bakermans-Kranenburg, M. J., M. H. van Ijezdoorn, and F. Juffer. 2008. Earlier is better: A meta-analysis of 70 years of intervention improving cognitive development in institutionalized children. *Monographs for Society in Research of Child Development* 73 (3): 279–293.

Banks, S. J., K. T. Eddy, M. Angstadt, P. J. Nathan, and K. L. Phan. 2007. Amygdala-frontal connectivity during emotion regulation. *Social and Affective Neuroscience* 2: 303–312.

Barker, D. J., G. Osmond, J. Golding, D. Kuh, and M. E. Wadsworth. 1989. Growth in utero, blood pressure in childhood and adult life and mortality from cardiovascular disease. *British Medical Journal* 298 (6673): 564–567.

Barker, D. J., P. D. Winter, C. Osmond, B. Margetts, S. J. Simmonds. 1989. Weight in infancy and death from ischaemic heart disease. *Lancet* 2 (8663): 577–580.

Barry, R. J., A. R. Clarke, and S. J. Johnstone. 2003. A review of electrophysiology in attention-deficit/hyperactivity disorder. I: Qualitative and quantitative electroencephalography. *Clinical Neurophysiology* 114: 171–183.

Barry, R. J., A. R. Clarke, S. J. Johnstone, R. McCarthy, and M. Selikowitz. 2009. Electroencephalogram theta/beta ratio and arousal in attention-deficit/hyperactivity disorder: Evidence of independent processes. *Biological Psychiatry* 66: 398–401.

Bauer, P. M., J. L. Hanson, R. K. Pierson, R. J. Davidson, and S. D. Pollak. 2009. Cerebellar volume and cognitive functioning in children who experienced early deprivation. *Biological Psychiatry* 66 (12): 1100–1106.

Bayley, N. 1969. *Bayley Scales of Infant Development.* New York: Psychological Corporation.

Bayley, N. 1993. *Bayley Scales of Infant Development,* 2nd ed. New York: Psychological Corporation.

Beckett, C., B. Maughan, M. Rutter, J. Castle, E. Colvert, C. Groothues, A. Hawkins, J. Kreppner, T. G. O'Connor, S. Stevens, and E. J. Sonuga-Barke. 2007. Scholastic attainment following severe early institutional deprivation: A study of children adopted from Romania. *Journal of Abnormal Child Psychology* 35: 1063–1073.

Beckett, C., B. Maughan, M. Rutter, J. Castle, C. Colvert, C. Groothues, J. Kreppner, S. Stevens, T. G. O'Connor, and E. J. Sonuga-Barke. 2006. Do the effects of early severe deprivation on cognition persist into early adolescence? Findings from the English and Romanian adoptees study. *Child Development* 77 (3): 696–711.

Beecher, H. K. 1966. Ethics and clinical research. *New England Journal of Medicine* 274: 1354–1360.

Belsky, J., M. J. Bakermans-Kranenburg, and M. H. van IJzendoorn. 2007. For better and for worse: Differential susceptibility to environmental influences. *Current Directions in Psychological Science* 6: 300–304.

Belsky, J., and M. Pluess. 2009. The nature (and nurture?) of plasticity in early human development. *Perspectives on Psychological Science* 4: 345–351.

Berument, S. K. 2013. Environmental enrichment and caregiver training to support the development of birth to 6-year-olds in Turkish orphanages. *Infant Mental Health Journal* 34: 189–201.

Binet, A., and T. Simon. 1904. Méthodes nouvelles pour le diagnostic du niveau intellectuel des anormaux. *L'Année psychologique* 11: 191–244.

Binet, A., and T. Simon. 1916. *The development of intelligence in children.* Baltimore: Williams and Wilkins; reprinted New York: Arno Press, 1973; Salem, NH: Ayer, 1983.

Black, J. E., T. A. Jones, C. A. Nelson, and W. T. Greenough. 1998. Neuronal plasticity and the developing brain. In *Handbook of child and adolescent psychiatry,* vol. 6: *Basic psychiatric science and treatment,* edited by N. E. Alessi, J. T. Coyle, S. I. Harrison, and S. Eth, 31–53. New York: John Wiley and Sons.

Blair, C., P. D. Zelazo, and M. T. Greenberg. 2005. The measurement of executive function in early childhood. *Developmental Neuropsychology* 28 (2): 561–571.

Blizzard, R. M., and A. Bulatovic. 1996. Syndromes of psychosocial short stature. In *Pediatric endocrinology,* edited by F. Lifshitz, 3rd ed., 83–93. New York: Marcel Dekker.

Boothby, N., R. L. Balster, P. Goldman, M. G. Wessells, C. H. Zeanah, G. Huebner, and J. Garbarino. 2012. Coordinated and evidence-based policy and practice for protecting children outside of family care. *Child Abuse and Neglect* 36: 743–751.

Bos, K. J., N. Fox, C. H. Zeanah, and C. A. Nelson. 2009. Effects of early psychosocial deprivation on the development of memory and executive function. *Frontiers in Behavioral Neuroscience,* vol. 3, art. 16: 1–7.

Bos, K., C. H. Zeanah, N. A. Fox, and C. A. Nelson. 2010. Stereotypies in children with a history of early institutional care. *Archives of Pediatrics and Adolescent Medicine* 164 (5): 406–411.

Boswell, J. 1988. *The kindness of strangers: The abandonment of children in western Europe from late antiquity to the Renaissance.* New York: Pantheon Books.

Bowlby, J. 1944. Forty-four juvenile thieves: Their characters and home-life. *International Journal of Psychoanalysis* 25: 19–52, 107–128.

Bowlby, J. 1951. Maternal care and mental health. *Bulletin of the World Health Organization* 3: 355–534. Available at http://libdoc.who.int/publications/9241400021_part1.pdf (to p. 63) and http://libdoc.who.int/publications/9241400021_part2.pdf (pp. 64–194).

Bowlby, J. 1969. *Attachment.* New York: Basic Books.

Bowlby, J. 1973. *Separation.* New York: Basic Books.

Bowlby, J. 1980. *Loss.* New York: Basic Books.

Brestan, E. V., and S. M. Eyberg. 1998. Effective psychosocial treatments for children and adolescents with disruptive behavior disorders: 29 years, 82 studies, and 5272 kids. *Journal of Clinical Child Psychology* 27: 179–188.

Brodbeck, A. J., and O. C. Irwin. 1946. The speech behaviour of infants without families. *Child Development* 17: 145–156.

Brodzinsky, D., and M. Schechter. 1990. *The psychology of adoption.* New York: Oxford University Press.

Browne, K., and C. Hamilton-Giachritsis. 2004. Mapping the number and characteristics of children under three in institutions across Europe at risk of harm. European Union, Daphne Programme—Year 2002: Final report, project no. 2002/017/C. Available at http://ec.europa.eu/justice_home/daphnetoolkit/files/projects/2002_017/01_final_report_2002_017.doc.

Browne, K. D., C. E. Hamilton-Giachritsis, R. Johnson, and M. Ostergren. 2006. Overuse of institutional care for children in Europe. *British Medical Journal* 332: 485–487.

Bruer, J. T. 1998. The brain and child development: Time for some critical thinking. *Public Health Reports* 113 (5): 388–397.

Bruininks, R. H., and B. D. Bruininks. 2005. *Bruininks-Oseretsky Test of Motor Proficiency,* 2nd ed. Minneapolis: NCS Pearson.

Bunge, S. A., and S. B. Wright. 2007. Neurodevelopmental changes in working memory and cognitive control. *Current Opinion in Neurobiology* 17 (2): 243–250.

Bunge, S. A., and P. D. Zelazo. 2006. A brain-based account of the development of rule use in childhood. *Current Directions in Psychological Science* 15 (3): 118–121.

Carlson, V., D. Cicchetti, D. Barnett, and K. Braunwald. 1989. Disorganized/disoriented attachment relationships in maltreated infants. *Developmental Psychology* 25: 525–531.

Carter, A. S., M. J. Briggs-Gowan, S. M. Jones, and T. D. Little. 2003. The

Infant-Toddler Social and Emotional Assessment (ITSEA): Factor structure, reliability, and validity. *Journal of Abnormal Child Psychology* 31: 495–514.

Casey, B. J., F. X. Castellanos, J. N. Giedd, et al. 1997. Implication of right frontostriatal circuitry in response inhibition and attention-deficit/hyperactivity disorder. *Journal of the American Academy of Child and Adolescent Psychiatry* 36: 374–383.

Casey, B. J., R. J. Trainor, J. L. Orendi, et al. 1997. A developmental functional MRI study of prefrontal activation during performance of a Go-No-Go task. *Journal of Cognitive Neuroscience* 9 (6): 835–847.

Cassidy, J., R. S. Marvin, and the MacArthur Working Group on Attachment. 1992. Attachment organization in preschool children: Coding guidelines, 4th ed. Unpublished manuscript, University of Virginia.

Castellanos, F. X., J. N. Giedd, W. L. Marsh, et al. 1996. Quantitative brain magnetic resonance imaging in attention-deficit hyperactivity disorder. *Archives of General Psychiatry* 53: 607–616.

Cermak, S. A., and L. A. Daunhauer. 1997. Sensory processing in the post-institutionalized child. *American Journal of Occupational Therapy* 51: 500–507.

Champagne, D. L., R. C. Bagot, F. van Hasselt, G. Ramakers, M. J. Meaney, E. R de Kloet, M. Joëls, and H. Krugers. 2008. Maternal care and hippocampal plasticity: Evidence for experience-dependent structural plasticity, altered synaptic functioning, and differential responsiveness to glucocorticoids and stress. *Journal of Neuroscience* 28 (23): 6037–6045.

Chapin, H. D. 1915. Are institutions necessary for infants? *Journal of the American Medical Association* 64: 1–3.

Chapin, H. D. 1926. Family vs. institution. *Survey* 55: 485–488.

Chapman, D. P., C. L. Whitfield, V. J. Felitti, S. R. Dube, V. J. Edwards, and R. F. Anda. 2004. Adverse childhood experiences and the risk of depressive disorders in adulthood. *Journal of Affective Disorders* 82: 217–225.

Children's Bureau. 2009. *Child Maltreatment 2009*. Washington, DC: U.S. Department of Health and Human Services, Administration for Children and Families, Administration on Children, Youth and Families, Children's Bureau. Available at http://archive.acf.hhs.gov/programs/cb/pubs/cm09/cm09.pdf.

Child Welfare League of America. 2007. Children in group homes and institutions by age and state, 2004. *Special Tabulation of the Adoption and Foster Care Analysis Reporting System*. Washington, DC: Author.

Chisholm, K. 1998. A three-year follow-up of attachment and indiscriminate friendliness in children adopted from Romanian orphanages. *Child Development* 69: 1092–1106.

Chisholm, K., M. C. Carter, E. W. Ames, and S. J. Morison. 1995. Attachment security and indiscriminately friendly behavior in children adopted from Romanian orphanages. *Development and Psychopathology* 7: 283–294.

Chou, S., and K. Browne. 2008. The relationship between institutional care and the international adoption of children in Europe. *Adoption and Fostering* 32: 40–48.

Chugani, H. T., M. E. Behen, O. Musik, C. Juhasz, F. Nagy, and D. C. Chugani. 2001. Local brain functional activity following early deprivation: A study of postinstitutionalized Romanian orphans. *NeuroImage* 14: 1290–1301.

Coles, E. M. 2001. Note on objectives of psychological assessment procedures and the value of multidimensional questionnaires and inventories. *Psychological Reports* 88: 813–816.

Colvert, E., M. Rutter, C. Beckett, J. Castle, C. Groothues, A. Hawkins, J. Kreppner, T. J. O'Connor, S. Stevens, and E. J. Sonuga-Barke. 2008. Emotional difficulties in early adolescence following severe early deprivation: Findings from the English and Romanian adoptees study. *Development and Psychopathology* 20: 547–567.

Colvert, E., M. Rutter, J. Kreppner, C. Beckett, J. Castle, C. Groothues, A. Hawkins, S. Stevens, and E. J. Sonuga-Barke. 2008. Do theory of mind and executive function deficits underlie the adverse outcomes associated with profound early deprivation? Findings from the English and Romanian adoptees study. *Journal of Abnormal Child Psychology* 36 (7): 1057–1068.

Council for International Organizations of Medical Sciences (CIOMS) in collaboration with the World Health Organization. 2002. *International ethical guidelines for biomedical research involving human subjects.* Geneva: CIOMS. Available at http://www.cioms.ch/publications/layout_guide2002.pdf.

Crăciun, O. 2011, January 18. Abandonați și de părinți, și de mamele sociale. Available at http://www.evz.ro/detalii/stiri/abandonati-si-de-parinti-si-de-mamele-sociale-918471.html#ixzz1kn9lJXmz EVZ.ro.

Cragg, L., and K. Nation. 2008. Go or no-go? Developmental improvements in the efficiency of response inhibition in mid-childhood. *Developmental Science* 11 (6): 819–827.

Crenson, M. A. 1998. *Building the invisible orphanage: A prehistory of the American welfare system.* Cambridge, MA: Harvard University Press.

Curtis, W. J., L. L. Lindeke, M. K. Georgieff, and C. A. Nelson. 2002. Neurobehavioral functioning in neonatal intensive care unit graduates in late childhood and early adolescence. *Brain* 125 (7): 1646–1659.

Cyr, C., E. M. Euser, M. J. Bakermans-Kranenburg, and M. H van IJzendoorn. 2010. Attachment security and disorganization in maltreating and high-risk families: A series of meta-analyses. *Development and Psychopathology* 22: 87–108.

Damjanovic, A. K., Y. Yang, R. Glaser, J. K. Keicolt-Glaser, H. Nguyen, B. Laskowski, Y. Zou, D. Q. Beversdorf, and N. P. Weng. 2007. Accelerated telomere erosion is associated with a declining immune function of caregivers of Alzheimer's disease patients. *Journal of Immunology* 179 (6): 4249–4254.

Davidson, R. J. 2000. Affective style, psychopathology, and resilience: Brain mechanisms and plasticity. *American Psychologist* 55: 1196–1214.

Davidson, R. J., and N. A. Fox. 1982. Asymmetrical brain activity discriminates between positive and negative stimuli in human infants. *Science* 218: 1235–1237.

de Haan, M. 2007. *Infant EEG and event-related potentials.* London: Psychology Press.

de Haan, M., K. Humphreys, and M. H. Johnson. 2002. Developing a brain specialized for face perception: A converging methods approach. *Developmental Psychobiology* 40 (3): 200–212.

de Haan, M., and C. A. Nelson. 1997. Recognition of the mother's face by 6-month-old infants: A neurobehavioral study. *Child Development* 68 (2): 187–210.

DeKlyen, M., and M. T. Greenberg. 2008. Attachment and psychopathology in childhood. In *Handbook of attachment: Theory, research, and clinical applications,* 2nd ed., edited by J. Cassidy and P. Shaver, 637–665. New York: Guilford.

Dennis, W., and P. Najarian. 1957. Infant development under environmental handicap. *Psychological Monographs: General and Applied* 71: 1–13.

Dente, K., and J. Hess. 2006. Pediatric AIDS in Romania: A country faces its epidemic and serves as a model of success. *Medscape General Medicine* 8 (2): 11.

De Wolff, M. S., and M. H. van IJzendoorn. 1997. Sensitivity and attachment: A meta-analysis on parental antecedents of infant attachment. *Child Development* 68: 571–591.

Dickens, J., and V. Groza. 2004. Empowerment in difficulty: A critical appraisal of international intervention in child welfare in Romania. *International Social Work Journal* 47: 469–487.

Dobrova-Krol, N. A., M. J. Bakermans-Kranenburg, M. H. van IJzendoorn, and F. Juffer. 2010. The importance of quality of care: Effects of perinatal HIV infection and early institutional rearing on preschoolers' attachment and indiscriminate friendliness. *Journal of Child Psychology and Psychiatry* 51: 1368–1376.

Dobrova-Krol, N. A., M. H. van IJzendoorn, M. J. Bakermans-Kranenburg, and F. Juffer. 2010. Effects of perinatal HIV infection and early institutional rearing on physical and cognitive development of children in Ukraine. *Child Development* 81 (1): 237–251.

Doolittle, T. 1995. The long term effects of institutionalization on the behavior of children from Eastern Europe and the former Soviet Union: Research, diagnoses, and therapy options. Parent Network for Post-Institutionalized Children. Retrieved July 17, 2004, from http://www.mariaschildren.org/english/ babyhouse/ effects.html.

Dozier, M., E. Peloso, O. Lindhiem, M. K. Gordon, M. Manni, S. Sepulveda, J. Ackerman, A. Bernier, and S. Levine. 2006. Preliminary evidence from a randomized clinical trial: Intervention effects on foster children's behavioral and biological regulation. *Journal of Social Issues* 62: 767–785.

Drury, S. S., K. P. Theal, A. T. Smyke, B. J. B. Keats, H. L. Egger, C. A. Nelson, N. A. Fox, P. J. Marshall, and C. H. Zeanah. 2010. Modification of depression by COMT val[158]met polymorphism in children exposed to early psychosocial deprivation. *Child Abuse and Neglect* 34: 387–395.

Drury, S. S., K. Theall, M. M. Gleason, A. T. Smyke, I. De Vivo, J. Y. Y. Wong, N. A. Fox, C. H. Zeanah, and C. A. Nelson. 2011. Telomere length and early severe social deprivation: Linking early adversity and cellular aging. *Molecular Psychiatry* 17 (7): 719–727.

Dube, S. R., R. F. Anda, V. J. Felitti, D. P. Chapman, D. F. Williamson, and W. H. Giles. 2001. Childhood abuse, household dysfunction, and the risk of attempted suicide throughout the life span: Findings from the adverse childhood experiences study. *Journal of the American Medical Association* 286: 3089–3096.

Durston, S., K. M. Thomas, M. S. Worden, Y. Yang, and B. J. Casey. 2002. The effect of preceding context on inhibition: An event-related fMRI study. *Neuroimage* 16 (2): 449–453.

Early Child Care Research Network. 1997. The effects of infant child care on infant-mother attachment security: Results of the NICHD study of early child care. *Child Development* 68: 860–879.

Edwards, V. J., G. W. Holden, V. J. Felitti, and R. F. Anda. 2003. Relationship between multiple forms of childhood maltreatment and adult mental

Gindis, B. 2000. Language-related problems and remediation strategies for internationally adopted orphanage-raised children. In *International adoption: Challenges and opportunities,* 2nd ed., edited by T. Tepper, L. Hannon, and D. Sandstrom, 89–97. Meadow Lands, PA: Parent Network for the Post-Institutionalized Child.

Gleason, M. M., N. A. Fox, S. Drury, A. T. Smyke, H. L. Egger, C. A. Nelson, M. G. Gregas, and C. H. Zeanah. 2011. The validity of evidence-derived criteria for reactive attachment disorder: Indiscriminately social/disinhibited and emotionally withdrawn/inhibited types. *Journal of the American Academy of Child and Adolescent Psychiatry* 50: 216–231.

Gleason, M. M., N. A. Fox, S. S. Drury, A. Smyke, C. A. Nelson, and C. H. Zeanah. Forthcoming. Indiscriminate behaviors in young children with a history of institutional care. *Pediatrics.*

Gogos, J. A., M. Morgan, V. Luine, M. Santha, S. Ogawa, D. Pfaff, and M. Karayiorgou. 1998. Catechol-O-methyltransferase-deficient mice exhibit sexually dimorphic changes in catecholamine levels and behavior. *Proceedings of the National Academy of Sciences* 95: 9991–9996.

Goldfarb, W. 1943. Infant rearing and problem behavior. *American Journal of Orthopsychiatry* 13: 249–265.

Goldfarb, W. 1944. Infant rearing as a factor in foster home replacement. *American Journal of Orthopsychiatry* 14: 162–173.

Goldfarb, W. 1945a. Effects of psychological deprivation in infancy and subsequent stimulation. *American Journal of Psychiatry* 102: 18–33.

Goldfarb, W. 1945b. Psychological privation in infancy and subsequent adjustment. *American Journal of Orthopsychiatry* 15: 247–255.

Goldsmith, H. H., and M. K. Rothbart. 1999. *The Laboratory Temperament Assessment Battery* (Locomotor version 3.1). Madison: University of Wisconsin.

Green, J., and R. Goldwyn. 2002. Attachment disorganization and psychopathology: New findings and their potential implications for developmental psychopathology in childhood. *Journal of Child Psychology and Psychiatry* 43: 679–690.

Greenough, W. T., J. E. Black, and C. S. Wallace. 1987. Experience and brain development. *Child Development* 58 (3): 539–559.

Greenwell, K. 2003. The effects of child welfare reform on levels of child abandonment and deinstitutionalization in Romania, 1987–2000. Ph.D. diss., University of Texas.

Gresham, F., and S. Elliot. 1990. *Social Skills Rating System Manual.* Minneapolis: Pearson.

Groza, V., and the American and Romanian Research Teams, in cooperation with Romanian Children's Relief/Fundatia Inocenti. 2001, November. A study of Romanian foster families. Available at http://msass. case.edu/downloads/vgroza/Final_Report_English.pdf.

Groza, V. K., K. M. Bunkers, and G. N. Gamer. 2011. Children without permanent parents: Research, practice and policy. VII: Ideal components and current characteristics of alternative care options for children outside of parental care in low-resource countries. *Monographs of Research in Child Development* 76 (4): 163–189.

Gunnar, M. R., and P. A. Fisher. 2006. Bringing basic research on early experience and stress neurobiology to bear on preventive interventions for neglected and maltreated children. *Developmental Psychopathology* 18 (3): 651–677.

Gunnar, M. R., and D. A. Kertes. 2003. Early risk factors and development of internationally adopted children: Can we generalize from the Romanian case? Paper presented at the Society for Research in Child Development, Tampa, FL.

Hacsi, T. A. 1997. *Second home: Orphan asylums and poor families in America.* Cambridge, MA: Harvard University Press.

Hague Convention on Protection of Children and Co-operation in Respect of Intercountry Adoption. 1993. Available at www.hcch.net/index_en.php?act=conventions.pdfandcid=69.

Halperin, J. M., and K. P. Schulz. 2006. Revisiting the role of the prefrontal cortex in the pathophysiology of attention-deficit/hyperactivity disorder. *Psychological Bulletin* 132: 560–581.

Harden, B. J. 2002. Congregate care for infants and toddlers: Shedding new light on an old question. *Infant Mental Health Journal* 23: 476–495.

Harlow, H. F. 1958. The nature of love. *American Psychologist* 13: 673–685.

Harlow, H. F. 1962. The heterosexual affectional system. *American Psychologist* 17: 1–10.

Harlow, H., and S. Suomi. 1970. Nature of love: Simplified. *American Psychologist* 25: 161–168.

Harris, K. M., E. M. Mahone, and H. S. Singer. 2008. Nonautistic motor stereotypies: Clinical features and longitudinal follow-up. *Pediatric Neurology* 38 (4): 267–272.

Hart, B., and T. R. Risley. 1995. *Meaningful differences in the everyday experience of young American children.* Baltimore: Brookes.

Heller, S. S., A. T. Smyke, and N. W. Boris. 2002. Very young foster children and foster families: Clinical challenges and interventions. *Infant Mental Health Journal* 23: 555–575.

Hermann, M. J., A. C. Ehlis, H. Elgring, and A. J. Fallgatter. 2005. Early stages (P100) of face perception in humans as measured with event-related potentials (ERPs). *Journal of Neural Transmissions* 112 (8): 1073–1081.

Hill, M. N., and B. S. McEwen. 2010. Involvement of the endocannabinoid system in the neurobehavioural effects of stress and glucocorticoids. *Progress in Neuro-Psychopharmacology and Biological Psychiatry* 34 (5): 791–797.

Hodges, J., and B. Tizard. 1989a. IQ and behavioral adjustment of ex-institutional adolescents. *Journal of Child Psychology, Psychiatry, and Allied Disciplines* 30: 53–75.

Hodges, J., and B. Tizard. 1989b. Social and family relationships of ex-institutional adolescents. *Journal of Child Psychology, Psychiatry, and Allied Disciplines* 30: 77–97.

Hoeksbergen, R. A. C., J. ter Laak, C. van Dijkum, S. Rijk, K. Rijk, and F. Stoutjesdijk. 2003. Posttraumatic stress disorder in adopted children from Romania. *American Journal of Orthopsychiatry* 73: 255–265.

Hough, S. D. 2000. Risk factors for the speech and language development of children adopted from Eastern Europe and the former USSR. In *International adoption: Challenges and opportunities,* 2nd ed., edited by T. Tepper, L. Hannon, and D. Sandstrom, 99–119. Meadow Lands, PA: Parent Network for the Post-Institutionalized Child.

Human Rights Watch. 1999. *Human Rights Watch world report 1999: Events of December 1997–November 1998,* 420. New York: Human Rights Watch.

Human Rights Watch. 2004. *Children's rights: Orphans and abandoned children.* Retrieved July 13, 2004, from http://www.hrw.org/topic/childrens-rights/orphans-and-abandoned-children.

Hunt, J. M., K. Mohandessi, M. Ghodssi, and M. Akiyama. 1976. The psychological development of orphanage-reared infants: Interventions with outcomes. *Genetic Psychology Monographs* 94: 177–226.

Hunt, K. 1991, March 24. The Romanian baby bazaar. *New York Times,* Magazine. Available at http://www.nytimes.com/1991/03/24/magazine/the-romanian-baby-bazaar.html?pagewanted=allandsrc=pm.

Huttenlocher, P. R. 1979. Synaptic density in human frontal cortex: Developmental changes and effects of aging. *Brain Research* 163 (2): 195–205.

Huttenlocher, P. R. 1984. Synapse elimination and plasticity in developing human cerebral cortex. *American Journal of Mental Deficiency* 88 (5): 488–496.

Huttenlocher, P. R. 2002. *Neural plasticity: The effects of environment on the development of the cerebral cortex*. Cambridge, MA: Harvard University Press.

Huttenlocher, P. R., and A. S. Dabholkar. 1997. Regional differences in synaptogenesis in human cerebral cortex. *Journal of Comparative Neurology* 387 (2): 167–178.

Huttenlocher, P. R., and C. de Courten. 1987. The development of synapses in striate cortex of man. *Human Neurobiology* 6 (1): 1–9.

Hwang, H. J., K. H. Kim, Y. J. Jung, D. W. Kim, Y. H. Lee, and C. H. Im. 2011. An EEG-based real time cortical functional connectivity imaging system. *Medical and Biological Engineering and Computing* 49 (9): 985–995.

IRB Procedures. 1991, June 18. Office for Protection from Research Risks, United States Department of Health and Human Services. Available at http://www.csc.vsc.edu/researchguidelines/procedures.pdf.

Jarriel, T. 1990, October 5. Inside Romanian orphanages. Produced by J. Tomlin. *20/20*, ABC.

Jeon, H., C. A. Nelson, and M. Moulson. 2010. The effects of early experience on emotion recognition: A study of institutionalized children in Romania. *Infancy* 15 (2): 209–221.

Johnson, D. 2011. Children without permanent parents: Research practice, and policy. IV: Growth failure in institutionalized children. *Monographs of the Society for Research in Child Development* 76 (4): 92–106.

Johnson, D. E., D. Guthrie, A. T. Smyke, S. F. Koga, N. A. Fox, C. H. Zeanah, and C. A. Nelson. 2010. Growth and relations between auxology, caregiving environment and cognition in socially deprived Romanian children randomized to foster vs. ongoing institutional care. *Archives of Pediatrics and Adolescent Medicine* 164 (6): 507–516.

Johnson, D. E., L. C. Miller, S. Iverson, W. Thomas, B. Franchino, K. Dole, M. T. Kiernan, M. K. Georgieff, and M. K. Hostetter. 1992. The health of children adopted from Romania. *Journal of the American Medical Association* 268 (24): 3446–3451.

Johnson, R., K. Browne, and C. Hamilton-Giachritsis. 2006. Young children in institutional care at risk of harm. *Trauma, Violence, and Abuse* 7: 34–59.

Juffer, F., M. J. Bakermans-Kranenburg, and M. H. van IJzendoorn. 2005. The importance of parenting in the development of disorganized attachment: Evidence from a preventive intervention study in adoptive families. *Journal of Child Psychology and Psychiatry* 46: 263–274.

Kaler, S. R., and B. J. Freeman. 1994. Analysis of environmental deprivation: Cognitive and social development in Romanian orphans. *Journal of Child Psychiatry and Psychology* 35: 769–781.

Kananen, L., I. Surakka, S. Pirkola, J. Suvisaari, J. Lönnqvist, L. Peltonen, S. Ripatti, and I. Hovatta. 2010. Childhood adversities are associated with shorter telomere length at adult age both in individuals with an anxiety disorder and controls. *PLoS ONE* 5 (5): e10826.

Karoum, F., S. J. Chrapusta, and M. F. Egan. 1994. 3-Methoxytyramine is the major metabolite of released dopamine in the rat frontal cortex: Reassessment of the effects of antipsychotics on the dynamics of dopamine release and metabolism in the frontal cortex, nucleus accumbens, and striatum by a simple two pool model. *Journal of Neurochemistry* 63: 972–979.

Karoum, F., M. F. Egan, and R. J. Wyatt. 1994. Selective reduction in dopamine turnover in the rat frontal cortex and hypothalamus during withdrawal from repeated cocaine exposure. *European Journal of Pharmacology* 254: 127–132.

Kashy, D. A., and D. A. Kenny. 1999. The analysis of data from dyads and groups. In *Handbook of research methods in social psychology,* edited by H.T. Reis and C. M. Judd. New York: Cambridge University Press.

Kenny, D. A. 1996. Models of non-independence in dyadic research. *Journal of Social and Personal Relationships* 13: 279–294.

Kenny, D. A., and D. A. Kashy. 2010. Dyadic data analysis using multilevel modeling. In *The handbook of advanced multilevel analysis,* edited by J. Hox and J. K. Roberts. London: Taylor and Francis.

Kligman, G. 1998. *The politics of duplicity: Controlling reproduction in Ceausescu's Romania.* Berkeley: University of California Press.

Kline, M. W., S. Rugina, M. Ilie, R. F. Matusa, A. M. Schweitzer, N. R. Calles, and H. L. Schwarzwald. 2007. Long-term follow-up of 414 HIV-infected Romanian children and adolescents receiving lopinavir/ritonavir-containing highly active antiretroviral therapy. *Pediatrics* 119: e1116–1120.

Knox, N. 2004, June 15. Romania to ban international adoptions permanently. *USA Today.* Available at http://usatoday30.usatoday.com/news/world/2004-06-15-romania-adoptions_x.htm.

Knudsen, E. 2004. Sensitive periods in the development of the brain and behavior. *Journal of Cognitive Neuroscience* 16: 1412–1425.

Knudsen, E. I., J. J. Heckman, J. L. Cameron, and J. P. Shonkoff. 2006. Economic, neurobiological, and behavioral perspectives on building America's future workforce. *Proceedings of the National Academy of Sciences* 103: 10155–10162.

Kochanska, G., K. C. Coy, T. L. Tjebkes, and S. J. Husarek. 1998. Individual differences in emotionality in infancy. *Child Development* 69 (2): 375–390.

Kochanska, G., K. T. Murray, and E. T. Harlan. 2000. Effortful control in early childhood: Continuity and change, antecedents, and implications for social development. *Developmental Psychology* 36: 220–232.

Konrad, K., S. Neufang, C. Hanisch, G. R. Fink, and B. Herpertz-Dahlmann. 2006. Dysfunctional attentional networks in children with attention deficit/hyperactivity disorder: Evidence from an event-related functional magnetic resonance imaging study. *Biological Psychiatry* 59: 643–651.

Kraemer, H. C., G. T. Wilson, C. Fairburn, and W. S. Agras. 2002. Mediators and moderators of treatment effects in randomized clinical trials. *Archives of General Psychiatry* 59: 877–883.

Kreppner, J. M., T. G. O'Connor, M. Rutter, and the English and Romanian Adoptees Study Team. 2001. Can inattention/overactivity be an institutional deprivation syndrome? *Journal of Abnormal Child Psychology* 29: 513–528.

Larrieu, J. A., and C. H. Zeanah. 1998. An intensive intervention for infants and toddlers in foster care. In *Custody: Child and adolescent psychiatric clinics of North America,* edited by K. Pruett and M. Pruett, 357–371. Philadelphia: Williams and Wilkins.

Larrieu, J. A., and C. H. Zeanah. 2003. Treating infant-parent relationships in context of maltreatment: An integrated, systems approach. In *Treatment of infant-parent relationship disturbances,* edited by A. Sameroff, S. McDonough, and K. Rosenblum, 243–264. New York: Guilford.

Lee, A., and B. L. Hankin. 2009. Insecure attachment, dysfunctional attitudes, and low self-esteem predicting prospective symptoms of depression and anxiety during adolescence. *Journal of Clinical Child and Adolescent Psychology* 38: 219–231.

Le Grand, R., C. J. Mondloch, D. Maurer, and H. P. Brent. 2001. Neuroperception: Early visual experience and face processing. *Nature* 410: 890.

Leiden Conference on the Development and Care of Children without Permanent Parents. 2012. The development and care of institutionally reared children. *Child Development Perspectives* 6: 174–180.

Leppänen, J. M., and C. A. Nelson. 2009. Tuning the developing brain to social signals of emotion. *Nature Reviews Neuroscience* 10 (1): 37–47.

Leppanen, J. M., and C. A. Nelson. 2013. The emergence of perceptual preferences for social signals of emotion. In *Navigating the social world*, edited by M. Banaji and S. Gelman. Oxford: Oxford University Press.

Leuner, B., E. R. Glasper, and E. Gould. 2010. Parenting and plasticity. *Trends in Neurosciences* 33: 465–473.

Levin, A. R., C. H. Zeanah, Jr., N. A. Fox, and C. A. Nelson. In press. Motor outcomes in children exposed to early psychosocial deprivation. *Journal of Pediatrics*.

Levy, R. J. 1947. Effects of institutional vs. boarding home care on a group of infants. *Journal of Personality* 15: 233–241.

Lie, N., and D. Murarasu. 2001. A follow-back of men and women who grew up in Romanian orphanages. *Journal of Preventative Medicine* 9 (2): 20–31.

Liston, C., B. S. McEwen, and B. J. Casey. 2009. Psychosocial stress reversibly disrupts prefrontal processing and attentional control. *Proceedings of the National Academy of Science* 106 (3): 912–917.

Loman, M. M., A. E. Johnson, A. Westerlund, S. D. Pollak, C. A. Nelson, and M. R. Gunnar. 2013. The effect of early deprivation on executive attention in middle childhood. *Journal of Child Psychology and Psychiatry* 54: 37–45.

Loman, M. M., K. L. Wiik, K. A. Frenn, S. D. Pollak, and M. R. Gunnar. 2009. Postinstitutionalized children's development: Growth, cognitive, and language outcomes. *Journal of Developmental and Behavioral Pediatrics* 30 (5): 426–434.

Loue, S. 2004. Ethical issues in research: A Romanian case study. *Revista Romana de Bioetica* 2: 16–27.

Lowrey, L. G. 1940. Personality distortion and early institutional care. *American Journal of Orthopsychiatry* 10: 576–585.

Luciana, M., and C. A. Nelson. 1998. The functional emergence of prefrontally guided working memory systems in four- to eight-year-old children. *Neuropsychologia* 36 (3): 273–293.

Luciana, M., and C. A. Nelson. 2002. Assessment of neuropsychological function through use of the Cambridge Neuropsychological Testing Automated Battery: Performance in 4- to 12-year-old children. *Developmental Neuropsychology* 22 (3): 595–624.

Lung, F. W., N. C. Chen, and B. C. Shu. 2007. Genetic pathway of major depressive disorder in shortening telomeric length. *Psychiatric Genetics* 17 (3): 195–199.

Lyons-Ruth, K., M. A. Easterbrooks, and C. D. Cibelli. 1997. Infant attachment strategies, infant mental lag, and maternal depressive symptoms:

Predictors of internalizing and externalizing problems at age 7. *Developmental Psychology,* 33: 681–692.

Mackenzie, E. J. Review of evidence regarding trauma system effectiveness resulting from panel studies. *Journal of Trauma,* 47(3 Suppl): S34–41.

Main, M., and J. Solomon. 1990. Procedures for identifying infants as disorganized/disoriented during the Ainsworth Strange Situation. In *Attachment in the preschool years: Theory, research and intervention,* edited by M. T. Greenberg, D. Cicchetti, and E. M. Cummings, 121–160. Chicago: University of Chicago Press.

Marcovitch, S., L. Cesaroni, W. Roberts, and C. Swanson. 1995. Romanian adoption: Parents' dreams, nightmares and realities. *Child Welfare* 74: 936–1032.

Marcovitch, S., S. Goldberg, A. Gold, J. Washington, C. Wasson, K. Krekewich, and M. Handley-Derry. 1997. Determinants of behavioral problems in Romanian children adopted in Ontario. *International Journal of Behavioral Development* 20: 17–31.

Marshall, P. J., Y. Bar-Haim, and N. A. Fox. 2002. Development of the EEG from 5 months to 4 years of age. *Clinical Neurophysiology* 113 (8): 1199–1208.

Marshall, P. J., N. A. Fox, and the Bucharest Early Intervention Project Core Group. 2004. A comparison of the electroencephalogram between institutionalized and community children in Romania. *Journal of Cognitive Neuroscience* 16 (8): 1327–1338.

Marshall, P. J., B. Reeb, N. A. Fox, C. A. Nelson, and C. H. Zeanah. 2008. Effects of early intervention on EEG power and coherence in previously institutionalized children in Romania. *Development and Psychopathology* 20: 861–880.

Martin-Ruiz, C., H. O. Dickinson, B. Keys, E. Rowan, R. A. Kenny, and T. von Zglinicki. 2006. Telomere length predicts poststroke mortality, dementia, and cognitive decline. *Annals of Neurology* 60 (2): 174–180.

Maurer, D., C. J. Mondloch, and T. L. Lewis. 2006. Sleeper effects. *Developmental Science* 10: 40–47.

McCartney, K., M. T. Owen, C. L. Booth, A. Clarke-Stewart, and D. L. Vandell. 2004. Testing a maternal attachment model of behavior problems in early childhood. *Journal of Child Psychology and Psychiatry* 45: 765–778.

McDermott, J. M., A. Westerlund, C. H. Zeanah, C. A. Nelson, and N. A. Fox. 2012. Early adversity and neural correlates of executive function: Implications for academic adjustment. *Developmental Cognitive Neuroscience* 2S: S59–S66.

McDermott M., D. L. Noordsy, and M. Traum. 2013. Neuroleptic malignant syndrome during multiple antipsychotic therapy. *Community Mental Health Journal* 49 (1): 45–46.

McEwen, B. 1998. Protective and damaging effects of stress mediators. *New England Journal of Medicine* 338 (3): 171–179.

McEwen, B. S. 2007. Physiology and neurobiology of stress and adaptation: Central role of the brain. *Physiological Reviews* 87 (3): 873–904.

McGoron, L., M. M. Gleason, A. T. Smyke, S. S. Drury, C. A. Nelson, N. A. Fox, and C. H. Zeanah. 2012. Recovering from early deprivation: Attachment mediates effects of caregiving on psychopathology. *Journal of the American Academy of Child and Adolescent Psychiatry* 51: 683–693.

McLaughlin, K. A., N. A. Fox, C. H. Zeanah, and C. A. Nelson. 2011. Adverse rearing environments and neural development in children: The development of frontal electroencephalogram asymmetry. *Biological Psychiatry* 70: 1008–1015.

McLaughlin, K. A., N. A. Fox, C. H. Zeanah, M. A. Sheridan, P. J. Marshall, and C. A. Nelson. 2010. Delayed maturation in brain activity explains the association between early environmental deprivation and symptoms of attention-deficit/hyperactivity disorder (ADHD). *Biological Psychiatry* 68 (4): 329–336.

McLaughlin, K. A., M. A. Sheridan, W. Winter, N. A. Fox, C. H. Zeanah, and C. A. Nelson. Forthcoming. Widespread reductions in cortical thickness following severe early-life deprivation: A neurodevelopmental pathway to ADHD. *Biological Psychiatry.*

McLaughlin, K. A., C. H. Zeanah, N. A. Fox, and C. A. Nelson. 2012. Attachment security as a mechanism linking foster care placement to improved mental health outcomes in previously institutionalized children. *Journal of the American Academy of Child and Adolescent Psychiatry* 53 (1): 46–55.

Meaney, M. J. 2001. Maternal care, gene expression, and the transmission of individual differences in stress reactivity across generations. *Annual Review of Neuroscience* 24: 1161–1192.

Mehta, M. A., N. I. Golembo, C. Nosarti, E. Colvert, A. Mota, S. C. Williams, M. Rutter, and E. J. Sonuga-Barke. 2009. Amygdala, hippocampal and corpus callosum size following severe early institutional deprivation: The English and Romanian adoptees study pilot. *Journal of Child Psychology and Psychiatry* 50 (8): 943–951.

Miller, E. K. 2000. The prefrontal cortex and cognitive control. *Nature Reviews Neuroscience* 1: 59–65.

Miller, F. G. 2009. The randomized controlled trial as a demonstration project: An ethical perspective. *American Journal of Psychiatry* 166: 743–745.

Miller, F. G., and H. Brody. 2003. A critique of clinical equipoise: Therapeutic misconception in the ethics of clinical trials. *Hastings Center Report* 33: 19–28.

Miller, L. C., and H. W. Hendrie. 2000. Health of children adopted from China. *Pediatrics* 105: e76–81.

Millum, J., and E. J. Emanuel. 2007. The ethics of international research with abandoned children. *Science* 318: 1874–1875.

Mitroi, L. 2012. From the orfanotrofia to the institutions for irrecoverables: Tracing the origins of institutional care for orphaned and abandoned children in Romania. B.A. Honors thesis, History of Science Department, Harvard University.

Moffitt, T. E., A. Caspi, and M. Rutter. 2006. Measured gene-environment interactions in psychopathology: Concepts, research strategies, and implications for research, intervention, and public understanding of genetics. *Perspectives on Psychological Science* 1: 5–27.

Monk, C. S., and C. A. Nelson. 2002. The effects of hydrocortisone on cognitive and neural functions: A behavioral and event-related potential investigation. *Neuropsychopharmacology* 26 (4): 505–519.

Mostofsky, S. H., K. L. Cooper, W. R. Kates, M. B. Denckla, and W. E. Kaufman. 2002. Smaller prefrontal and premotor volumes in boys with attention-deficit/hyperactivity disorder. *Biological Psychiatry* 52: 785–794.

Moulson, M. C., N. A. Fox, C. H. Zeanah, and C. A. Nelson. 2009. Early adverse experiences and the neurobiology of facial emotion processing. *Developmental Psychology* 45: 17–30.

Moulson, M. C., A. Westerlund, N. A. Fox, C. H. Zeanah, and C. A. Nelson. 2009. The effects of early experience on face recognition: An event-related potential study of institutionalized children in Romania. *Child Development* 80 (4): 1039–1056.

Najjar, S. S., A. K. Khachadurian, M. N. Ilbawi, and R. M. Blizzard. 1971. Dwarfism with elevated levels of plasma growth hormone. *New England Journal of Medicine* 284 (15): 809–812.

Narr, K. L., R. P. Woods, J. Lin, et al. 2009. Widespread cortical thinning is a robust anatomical marker for attention-deficit/hyperactivity disorder. *Journal of the American Academy of Child and Adolescent Psychiatry* 48: 1014–1022.

National Authority for Child Protection and Adoption. 2004. *Protectia co-*

pilului: Intre rezultate obtinute si prioritati pentru viitor. Bucharest: Government of Romania.

Nelson, C. A. 1997. The neurobiological basis of early memory development. In *The development of memory in childhood,* edited by N. Cowan, 41–82. Hove, East Sussex, UK: Psychology Press.

Nelson, C. A. 1998. The nature of early memory. *Preventive Medicine* 27: 172–179.

Nelson, C. A. 2001. The development and neural bases of face recognition. *Infant and Child Development* 10: 3–18.

Nelson, C. A. 2007. A neurobiological perspective on early human deprivation. *Child Development Perspectives* 1: 13–18.

Nelson, C. A., E. A. Furtado, N. A. Fox, and C. H. Zeanah. 2009. The deprived human brain. *American Scientist* 97: 222–229.

Nelson, C. A., and S. Jeste. 2008. Neurobiological perspectives on developmental psychopathology. In *Textbook on child and adolescent psychiatry,* 5th ed., edited by M. Rutter, D. Bishop, D. Pine, S. Scott, J. Stevenson, E. Taylor, and A. Thapar, 145–159. London: Blackwell.

Nelson, C. A., and M. Luciana, eds. 2008. *Handbook of developmental cognitive neuroscience,* 2nd ed. Cambridge, MA: MIT Press.

Nelson, C. A., and J. P. McCleery. 2008. The use of event-related potentials in the study of typical and atypical development. *Journal of the American Academy of Child and Adolescent Psychiatry* 47 (11): 1252–1261.

Nelson, C. A., and C. Monk. 2001. The use of event-related potentials in the study of cognitive development. In *Handbook of developmental cognitive neuroscience,* edited by C. A. Nelson and M. Luciana, 125–136. Cambridge, MA: MIT Press.

Nelson, C. A., S. W. Parker, D. Guthrie, and the Bucharest Early Intervention Project Core Group. 2006. The discrimination of facial expressions by typically developing infants and toddlers and those experiencing early institutional care. *Infant Behavior and Development* 29 (2): 210–219.

Nelson, C. A., A. Westerlund, J. M. McDermott, C. H. Zeanah, and N. A. Fox. 2013. Emotion recognition following early psychosocial deprivation. *Developmental Psychopathology* 25: 517–525.

Nelson, C. A., C. H. Zeanah, N. A. Fox, P. J. Marshall, A. T. Smyke, and D. Guthrie. 2007. Cognitive recovery in socially deprived young children: The Bucharest Early Intervention Project. *Science* 318 (5858): 1937–1940, and supplemental online material.

Netter, S., and Z. Magee. 2010, April 9. Tennessee mother ships adopted son back to Moscow alone. ABC World News with Diane Sawyer.

http://abcnews.go.com/WN/anger-mom-adopted-boy-back-russia/
story?id=10331728#.

NGO Working Group on Children without Parental Care. 2006. Development of international standards for the protection of children deprived of parental care. Child Rights, the Role of Families and Alternative Care: Policies Developments, Trends and Challenges in Europe.
International Conference, Bucharest, 2–3 February. Available at http://
www.crin.org/docs/Bucharest%20Conference%20Presentation.doc.

NICHD Early Child Care Research Network. 1996. Characteristics of
infant child care: Factors contributing to positive caregiving. *Early
Childhood Research Quarterly* 11: 269–306.

Nigg, J. T., and B. J. Casey. 2005. An integrative theory of attention-deficit/
hyperactivity disorder based on the cognitive and affective neurosciences. *Development and Psychopathology* 17: 785–806.

Noble, K. G., B. D. McCandliss, M. J. Farah. 2007. Socioeconomic gradi-
ents predict individual differences in neurocognitive abilities. *Developmental Science* 10 (4): 464–480.

Norris, C. L. 2009. The banning of international adoption in Romania:
Reasons, meaning and implications for child care and protection. Ph.D.
diss., Boston University.

Novotny, T., D. Haazen, and O. Adeyi. 2003. *HIV/AIDS in Southeastern
Europe: Case studies from Bulgaria, Croatia, and Romania.* Washington,
DC: World Bank.

O'Connor, T. G., D. Bredenkamp, M. Rutter, and the English and Romanian Adoptees Study Team. 1999. Attachment disturbances and disorders in children exposed to early and severe deprivation. *Infant Mental
Health Journal* 20: 10–29.

O'Connor, T. G., R. S. Marvin, M. Rutter, J. T. Olrick, P. A. Brittner, and
the English and Romanian Adoptees (ERA) Study Team. 2003. Child-
parent attachment following severe early institutional deprivation. *Development and Psychopathology* 15: 19–38.

O'Connor, T. G., M. Rutter, and the English and Romanian Adoptees
Study Team. 2000. Attachment disorder behavior following early severe
deprivation: Extension and longitudinal follow-up. *Journal of the American Academy of Child and Adolescent Psychiatry* 39: 703–712.

Ollendick, T. H., M. D. Weist, M. C. Borden, and R. W. Greene. 1992. So-
ciometric status and academic, behavioral, and psychological adjustment: A five-year longitudinal study. *Journal of Consulting and Clinical
Psychology* 60: 80–87.

Parker, S. W., C. A. Nelson, and the Bucharest Early Intervention Project

Core Group. 2005a. The impact of early institutional rearing on the ability to discriminate facial expressions of emotion: An event-related potential study. *Child Development* 76 (1): 54–72.

Parker, S. W., C. A. Nelson, and the Bucharest Early Intervention Project Core Group. 2005b. An event-related potential study of the impact of institutional rearing on face recognition. *Development and Psychopathology* 17: 621–639.

Parker, J., K. Rubin, X. Erath, J. Wojslawowicz, and A. Buskirk. 2006. Developmental psychopathology: Risk, disorder, and adaptation. In *Developmental psychopathology,* 2nd ed., edited by D. Cicchetti and D. Cohen, 419–493. New York: John Wiley and Sons.

Parks, C. G., D. B. Miller, E. C. McCanlies, R. M. Cawthon, M. E. Andrew, L. A. DeRoo, and D. P. Sandler. 2009. Telomere length, current perceived stress, and urinary stress hormones in women. *Cancer Epidemiology Biomarkers and Prevention* 18 (2): 551–560.

Participants in the 2001 Conference on Ethical Aspects of Research in Developing Countries. 2002. Fair benefits for research in developing countries. *Science* 298 (5601): 2133–2134.

Phillips, N., C. Hammen, P. Brennan, J. Najman, and W. Bor. 2005. Early adversity and the prospective prediction of depressive and anxiety disorders in adolescents. *Journal of Abnormal Child Psychology* 33 (1): 13–24.

Pinheiro, P. S. 2006. *World report on violence against children.* Geneva: United Nations. Available at http://www.unicef.org/lac/full_tex(3).pdf.

Pollak, S. D. 2005. Early adversity and mechanisms of plasticity: Integrating affective neuroscience and developmental approaches to psychopathology. *Developmental Psychopathology* 17 (3): 725–752.

Pollak, S. D., L. L. Holt, and A. B. Wismer Fries. 2004. Hemispheric asymmetries in children's perception of nonlinguistic human affective sounds. *Developmental Science* 7 (1): 10–18.

Pollak, S. D., C. A. Nelson, M. F. Schlaak, B. J. Roeber, S. S. Wewerka, K. L. Wiik, K. A. Frenn, M. M. Loman, and M. R. Gunnar. 2010. Neurodevelopmental effects of early deprivation in postinstitutionalized children. *Child Development* 81 (1): 224–236.

Pollak, S. D., and P. Sinha. 2002. Effects of early experience on children's recognition of facial displays of emotion. *Developmental Psychology* 38: 784–791.

Popa-Mabe, M. C. 2010. "Ceausescu's orphans": Narrating the crisis of Romanian international child adoption. Ph.D. diss., Bryn Mawr College. Available from ProQuest Dissertations and Theses database, UMI no. 3402968.

Powell, G. E., J. A. Brasel, and R. M. Blizzard. 1967. Emotional deprivation and growth retardation simulating idiopathic hypopituitarism. I. Clinical evaluation of the syndrome. *New England Journal of Medicine* 276: 1271–1278.

Provence, S., and R. C. Lipton. Preface by M. J. E. Senn. 1962. *Infants in institutions: A comparison of their development with family-reared infants during the first year of life.* New York: International Universities Press.

Puscasu, G., and B. Codres. 2011. Nonlinear system identification and control based on modular neural networks. *International Journal of Neural Systems* 21 (4): 319–334.

Ramey, C. T., and F. A. Campbell. 1984. Preventive education for high-risk children: Cognitive consequences of the Carolina Abecedarian Project. *American Journal of Mental Deficiency* 88 (5): 515–523.

Rauscher, F. H., G. L. Shaw, and K. N. Ky. 1993. Music and spatial task performance. *Nature* 365: 611.

Rheingold, H. L. 1961. The effect of environmental stimulation upon social and exploratory behavior in the human infant. In *Determinants of infant behavior,* edited by B. M. Foss, 143–177. London: Methuen.

Rid, A. 2012. When is research socially valuable? Lessons from the Bucharest Early Intervention Project. *Journal of Nervous and Mental Disease* 200: 248–249.

Righi, G., and C. A. Nelson. 2013. The neural architecture and developmental course of face processing. In *Neural circuit development and function in the brain,* edited by J. Rubenstein and P. Rakic, 331–350. San Diego: Academic Press.

Robertson, J., and J. Robertson. 1989. *Separation and the very young.* London: Free Association Books.

Roger, C., C. G. Bénar, F. Vidal, T. Hasbroucq, and B. Burle. 2010. Rostral cingulate zone and correct response monitoring: ICA and source localization evidences for the unicity of correct- and error-negativities. *NeuroImage* 51(1): 391–403.

Rosapepe, J. C. 2001. *Half way home: Romania's abandoned children ten years after the revolution.* Report to Americans from the U.S. Embassy, Bucharest, Romania.

Rosenberg, D. R., K. Pajer, and M. Rancurello. 1992. Neuropsychiatric assessment of orphans in one Romanian orphanage for "unsalvageables." *Journal of the American Medical Association* 268 (24): 3489–3490.

Rousseau, J.-J. (1781) 1953. *The confessions.* Harmondsworth, UK: Penguin Books.

Rousseau, J.-J. (1762) 1980. *Emile.* Meppel: Boom.

Roy, Chaitali B. 2010, October 4. Child trafficking new form of slavery. *Arab Times.* Available at http://www.arabtimesonline.com/NewsDetails/tabid/96/smid/414/ArticleID/160266/reftab/36/t/Child-trafficking-new-form-of-slavery/Default.aspx.

Roy, P., M. Rutter, and A. Pickles. 2000. Institutional care: Risk from family background or pattern of rearing? *Journal of Child Psychology and Psychiatry* 41: 139–141.

Roy, P., M. Rutter, and A. Pickles. 2004. Institutional care: Associations between over-activity and a lack of selectivity in social relationships. *Journal of Child Psychology and Psychiatry* 45: 866–873.

Rubia, K., S. Overmeyer, E. Taylor, et al. 1999. Hypofrontality in attention deficit hyperactivity disorder during higher-order motor control: A study with functional MRI. *American Journal of Psychiatry* 156: 891–896.

Rutter, M. 1996. Connections between child and adult psychopathology. *European Child and Adolescent Psychiatry* 5 Suppl 1: 4–7.

Rutter, M., L. Andersen-Wood, C. Beckett, D. Bredenkamp, J. Castle, C. Groothues, J. Kreppner, L. Keaveney, C. Lord, T. G. O'Connor, and the English and Romanian Adoptees Study Team. 1999. Quasi-autistic patterns following severe early global privation. *Journal of Child Psychology and Psychiatry* 40: 537–549.

Rutter, M., E. Colvert, J. Kreppner, C. Beckett, J. Castle, C. Groothues, A. Hawkins, T. G. O'Connor, S. E. Stevens, E. J. Sonuga-Barke. 2007. Early adolescent outcomes for institutionally-deprived and nondeprived adoptees. I: Disinhibited attachment. *Journal of Child Psychology and Psychiatry* 48: 17–30.

Rutter, M., and the English and Romanian Adoptees Study Team. 1998. Developmental catch-up, and deficit, following adoption after severe global early privation. *Journal of Child Psychology and Psychiatry* 39: 465–476.

Rutter, M., J. M. Kreppner, and T. O'Connor. 2001. Specificity and heterogeneity in children's responses to profound institutional privation. *British Journal of Psychiatry* 179: 97–103.

Rutter, M., J. Kreppner, and E. Sonuga-Barke. 2009. Emanuel Miller Lecture: Attachment insecurity, disinhibited attachment, and attachment disorders: Where do research findings leave the concepts? *Journal of Child Psychology and Psychiatry* 50: 529–543.

Rutter, M., T. G. O'Connor, and the English and Romanian Adoptees (ERA) Study Team. 2004. Are there biological programming effects for psychological development? Findings from a study of Romanian adoptees. *Developmental Psychology* 40: 81–94.

Rutter, M., and E. J. Sonuga-Barke. 2010. X. Conclusions: Overview of findings from the era study, inferences and research implications. *Monographs of the Society for Research in Child Development* 75 (1): 212–229.

Rutter, M., E. J. Sonuga-Barke, C. Beckett, J. Castle, J. Kreppner, R. Kumsta, W. Schlotz, S. E. Stevens, C. A. Bell, and M. R. Gunnar. 2010. Deprivation-specific psychological patterns: Effects of institutional deprivation. *Monographs of the Society for Research in Child Development* 75 (1, serial no. 295).

Sanchez, M. M., C. O. Ladd, and P. M. Plotsky. 2001. Early adverse experience as a developmental risk factor for later psychopathology: Evidence from rodent and primate models. *Developmental Psychopathology* (13) 3: 419–449.

Sandu, A. I. 2006. Poverty, institutions and child health in a post communist rural Romania: A view from below. Ph.D. diss., Department of Public Administration, Syracuse University. Available at http://surface.syr.edu/ppa_etd/11/.

Save the Children. 2009. Institutional care: The last resort. Policy brief. London. http://www.savethechildren.org.uk/resources/online-library/keeping-children-out-of-harmful-institutions-why-we-should-be-investing-in-family-based-care.

Scheffel, J. K. 2005. Beyond death towards solicitude in Bulgaria and Romania: A critical hermeneutic inquiry of home among orphaned and abandoned children. Ed.D. diss., School of Education, University of San Francisco. Available from ProQuest Dissertations and Theses database, UMI no. 3167896.

Shaw, P., K. Eckstrand, W. Sharp, et al. 2007. Attention-deficit/hyperactivity disorder is characterized by a delay in cortical maturation. *Proceedings of the National Academy of Sciences* 104: 19649–19654.

Shaw, P., J. P. Lerch, D. Greenstein, et al. 2006. Longitudinal mapping of cortical thickness and clinical outcome in children and adolescents with attention-deficit/hyperactivity disorder. *Archives of General Psychiatry* 63: 540–549.

Sheridan, M. A., N. A. Fox, C. H. Zeanah, K. McLaughlin, and C. A. Nelson. 2012. Variation in neural development as a result of exposure to institutionalization early in childhood. *Proceedings of the National Academy of Sciences* 109 (32): 12927–12932.

Shonkoff, J. P., and D. A. Phillips, eds. 2000. From neurons to neighborhoods: The science of early childhood development. Washington, DC: National Academy Press.

Simion, F., E. Di Giorgio, I. Leo, and L. Bardi. 2011. The processing of social stimuli in early infancy: From faces to biological motion perception. *Progress in Brain Research* 189: 173–193.

Simon, N. M., J. W. Smoller, K. L. McNamara, R. S. Maser, A. K. Zalta, M. H. Pollack, A. A. Nierenberg, M. Fava, and K. K. Wong. 2006. Telomere shortening and mood disorders: Preliminary support for chronic stress model of accelerated aging. *Biological Psychiatry* 60 (5): 432–435.

Skeels, H. M. 1966. Adult status of children with contrasting early life experiences: A follow up study. *Monographs of the Society for Research in Child Development* 31 (3, serial no. 105).

Skeels, H. M., and M. Skodak. 1965. Techniques for a high-yield follow-up study in the field. *Public Health Reports* 80: 249–257.

Slopen, N., K. A. McLaughlin, N. A. Fox, C. H. Zeanah, and C. A. Nelson. 2012. Alterations in neural processing and psychopathology in children raised in institutions. *Archives of General Psychiatry* 69: 1022–1030.

Sloutsky, V. M. 1997. Effects of institutional care on cognitive and social development of six- and seven-year-old children: A contextualist perspective. *International Journal of Behavioral Development* 20: 131–153.

Smith, M. G., and R. Fong. 2004. *The children of neglect.* New York: Brunner-Routledge.

Smyke, A. T., and A. S. Breidenstine. 2009. Foster care in early childhood. In *Handbook of infant mental health,* 3rd ed., edited by C. H. Zeanah, 500–515. New York: Guilford Press.

Smyke, A. T., A. Dumitrescu, and C. H. Zeanah. 2002. Attachment disturbances in young children. I: The continuum of caretaking casualty. *Journal of the American Academy of Child and Adolescent Psychiatry* 41 (8): 972–982.

Smyke, A. T., S. F. Koga, D. E. Johnson, N. A. Fox, P. J. Marshall, C. A. Nelson, C. H. Zeanah, and the Bucharest Early Intervention Project Core Group. 2007. The caregiving context in institution-reared and family-reared infants and toddlers in Romania. *Journal of Child Psychology and Psychiatry* 48 (2): 210–218.

Smyke, A. T., C. H. Zeanah, N. A. Fox, and C. A. Nelson. 2009a. A new model of foster care for young children: The Bucharest Early Intervention Project. *Child and Adolescent Psychiatry Clinics of North America* 18 (3): 721–734.

Smyke, A. T., C. H. Zeanah, N. A. Fox, and C. A. Nelson. 2009b. Psychosocial interventions: Bucharest Early Intervention Project. In *Infant and early childhood mental health,* edited by D. Schechter and M. M. Gleason, special issue, *Child and Adolescent Psychiatric Clinics of North America* 18: 721–734.

Smyke, A. T., C. H. Zeanah, N. A. Fox, C. A. Nelson, and D. Guthrie. 2010. Placement in foster care enhances quality of attachment among young institutionalized children. *Child Development* 81: 212–223.

Smyke, A. T., C. H. Zeanah, M. M. Gleason, S. S. Drury, N. A. Fox, C. A. Nelson, and D. G. Guthrie. 2012. A randomized controlled trial of foster care vs. institutional care for children with signs of reactive attachment disorder. *American Journal of Psychiatry* 169: 508–514.

Sonuga-Barke, E. J., W. Schlotz, and M. Rutter. 2008. Deprivation-specific psychological patterns. VII: Physical growth and maturation following early severe institutional deprivation: Do they mediate specific psychopathological effects? *Monographs of the Society for Research in Child Development* 75 (1): 143–166.

Sowell, E. R., P. M. Thompson, C. J. Holmes, T. L. Jernigan, and A. W. Toga. 1999. In vivo evidence for post-adolescent brain maturation in frontal and striatal regions. *Nature Neuroscience* 2 (10): 859–851.

Sparling, J., C. Dragomir, S. L. Ramey, and L. Florescu. 2005. An educational intervention improves developmental progress of young children in a Romanian orphanage. *Infant Mental Health Journal* 26 (2): 127–142.

Spitz, E. B., C. Brenner, and C. Davison. 1945. A new absorbable material for use in neurological and general surgery. *Science* 102 (2658): 621.

Spitz, R. 1945. Hospitalism: An inquiry into the genesis of psychiatric conditions in early childhood. *Psychoanalytic Study of the Child* 1: 53–74.

Spitz, R. 1952. *Psychogenic diseases in infancy: An attempt at their classification.* A film by the Psychoanalytic Research Project on Problems of Infancy. Available at http://www.youtube.com/watch?v=VvdOe10vrs4.

Stativă, E., C. Anghelescu, R. Mitulescu, M. Nanu, and N. Stanciu. 2005. *The situation of child abandonment in Romania.* Geneva: UNICEF. Available at http://www.ceecis.org/child_protection/PDF/child%20abandonment%20in%20Romania.pdf.

Stern, D. N. 1985. *The interpersonal world of the infant.* New York: Basic Books.

Stevens, S. E., E. J. Sonuga-Barke, J. M. Kreppner, C. Beckett, J. Castle, E. Colvert, C. Groothues, A. Hawkins, and M. Rutter. 2008. Inattention/overactivity following early severe institutional deprivation: Presentation and associations in early adolescence. *Journal of Abnormal Child Psychology* 36: 385–398.

St. Petersburg–USA Orphanage Research Team. 2008. The effects of early social-emotional and relationship experience on the development of young children. *Monographs of the Society for Research in Child Development* 72 (3, serial no. 291).

Stromswold, K. 1995. The cognitive and neural bases of language acquisition. In *The cognitive neurosciences,* edited by M. S. Gazzaniga, 855–870. Cambridge, MA: MIT Press.

Szyf, M., I. Weaver, and M. Meaney. 2007. Maternal care, the epigenome and phenotypic differences in behavior. *Reproductive Toxicology* 24 (1): 9–19.

Tarabulsy, G. M., R. Tessier, and A. Kappas. 1996. Contingency detection and the contingent organization of behavior interactions: Implications for socioemotional development in infancy. *Psychological Bulletin* 120: 25–41.

Terman, L. M., and M. A. Merrill. 1937. *Measuring intelligence: A guide to the administration of the new revised Stanford–Binet tests of intelligence.* Boston: Houghton Mifflin.

Thomas, L. A., M. D. De Bellis, R. Graham, and K. S. LaBar. 2007. Development of emotional facial recognition in late childhood and adolescence. *Developmental Science* 10 (5): 547–558.

Thompson, R. A., and C. A. Nelson. 2001. Developmental science and the media: Early brain development. *American Psychologist* 56: 5–15.

Tizard, B. 1977. *Adoption: A second chance.* London: Open Books.

Tizard, B., and J. Hodges. 1978. The effect of early institutional rearing on the development of eight-year-old children. *Journal of Child Psychology and Psychiatry* 19: 99–118.

Tizard, B., and J. Rees. 1974. A comparison of the effects of adoption, restoration to the natural mother, and continued institutionalization on the cognitive development of four-year-old children. *Child Development* 45: 92–99.

Tizard, B., and J. Rees. 1975. The effect of early institutional rearing on the behavior problems and affectional relationships of four-year-old children. *Journal of Child Psychology and Psychiatry* 16: 61–73.

Tobis, D. 2000. Moving from residential institutions to community-based social services in Central and Eastern Europe and the former Soviet Union. Washington, DC: World Bank.

Tottenham, N., T. A. Hare, A. Millner, T. Gilhooly, J. D. Zevin, and B. J. Casey. 2011. Elevated amygdala response to faces following early deprivation. *Developmental Science* 14 (2): 190–204.

Tottenham, N., T. A. Hare, B. T. Quinn, T. W. McCarry, M. Nurse, T. Gilhooly, and B. J. Casey. 2010. Prolonged institutional rearing is associated with atypically large amygdala volume and difficulties in emotion regulation. *Developmental Science* 13 (1): 46–61.

Tottenham, N., and M. A. Sheridan. 2009. A review of adversity, the amygdala and the hippocampus: A consideration of developmental timing. *Frontiers in Human Neuroscience* 3: 68.

Tyrka, A. R., L. H. Price, H. T. Kao, B. Porton, S. A. Marsella, and L. L. Carpenter. 2010. Childhood maltreatment and telomere shortening: Preliminary support for an effect on early stress on cellular aging. *Biological Psychiatry* 67 (6): 531–534.

United Nations. 1989, November 20. Convention on the rights of the child. Resolution 44/25. Available at http://www.un.org/documents/ga/res/44/a44r025.htm.

UNICEF. 2003. *Children in institutions: The beginning of the end? The cases of Italy, Spain, Argentina, Chile and Uruguay.* Florence, Italy: UNICEF Innocenti Research Centre. Available at http://www.unicef-irc.org/publications/pdf/insight8e.pdf.

United Nations Development Programme—Romania. 2002. *A decade later: Understanding the transition process in Romania.* National Human Development Report. Available at http://hdr.undp.org/en/reports/national/europethecis/romania/Romania_2001_en.pdf.

U.S. Department of Health and Human Services. 1979. The Belmont Report: Ethical guidelines for the protection of human subjects of biomedical and behavioral research. Washington, DC: U.S. Government Printing Office. Available at http://www.hhs.gov/ohrp/humansubjects/guidance/belmont.html.

U.S. Government. 2010. A whole-of-government approach to child welfare and protection. Fourth Annual Report to Congress on Public Law 109-95. Available at http://pdf.usaid.gov/pdf_docs/PDACQ777.pdf.

U.S. Government. 2012. United States Government action plan on children in adversity: A framework for international assistance; 2012–2017. Available at http://transition.usaid.gov/our_work/global_health/pdf/apca.pdf.

van den Dries, L., F. Juffer, M. H. van IJzendoorn, and M. J. Bakermans-Kranenburg. 2010. Infants' physical and cognitive development after international adoption from foster care or institutions in China. *Journal of Developmental and Behavioral Pediatrics* 31 (2): 144–150.

Vanderwert, R. E., P. J. Marshall, C. A. Nelson, C. H. Zeanah, and N. A. Fox. 2010. Timing of intervention affects brain electrical activity in children exposed to severe psychosocial neglect. *PLoS ONE* 5 (7): 1–5.

van IJzendoorn, M. H., S. Goldberg, P. M. Kroonenberg, and O. J. Frenkel. 1992. The relative effects of maternal and child problems on the quality of attachment: A meta-analysis of attachment in clinical samples. *Child Development* 63: 840–858.

van IJzendoorn, M. H., M. P. C. M. Luijk, and F. Juffer. 2008. IQ of children growing up in children's homes: A meta-analysis on IQ delays in orphanages. *Merrill Palmer Quarterly* 54: 341–366.

van IJzendoorn, M. H., C. Schuengel, and M. J. Bakermans-Kranenburg. 1999. Disorganized attachment in early childhood: Meta-analysis of precursors, concomitants, and sequelae. *Development and Psychopathology* 11: 225–249.

van Veen, V., and C. S. Carter. 2002. The anterior cingulate as a conflict monitor: fMRI and ERP studies. *Physiology and Behavior* 77 (4–5): 477–482.

Viazzo P. P., M. Bortolotto, and A. Zanotto. 2000. Five centuries of foundling history in Florence: Changing patterns of abandonment, care and mortality. In *Abandoned children,* edited by C. Panter-Brick and M. E. Smith, 70–91. Cambridge: Cambridge University Press.

Volkmar, F., C. Lord, A. Klin, R. Schultz, and E. Cook. 2007. Autism and the pervasive developmental disorders. In *Lewis' textbook of child and adolescent psychiatry: A comprehensive textbook,* edited by A. Martin and F. R. Volkmar, 384–400. Philadelphia: Lippincott Williams and Wilkins.

Vorria, P., Z. Papaligoura, J. Dunn, M. H. van IJzendoorn, H. Steele, A. Kontopoulou, and Y. Sarafidou. 2003. Early experiences and attachment relationships of Greek infants raised in residential group care. *Journal of Child Psychology and Psychiatry* 44: 1208–1220.

Wajda-Johnston, V., A. T. Smyke, G. Nagle, and J. A. Larrieu. 2005. Using technology as a training, supervision, and consultation aid. In *The handbook of training and practice in infant and preschool mental health,* edited by K. M. Finello, 357–374. San Francisco: John Wiley and Sons.

Walker, S. P., T. D. Wachs, S. Grantham-McGregor, M. M. Black, C. A. Nelson, S. L. Huffman, H. Baker-Henningham, S. M. Chang, J. D. Hamadani, B. Lozoff, J. M. Meeks Gardner, C. A. Powell, A. Rahman, and L. Richter. 2011. Inequality in early childhood: Risk and protective factors for early child development. *Lancet* 378: 1325–1338.

Warren, S., L. Huston, B. Egeland, and L. A. Sroufe. 1997. Child and adolescent anxiety disorders and early attachment. *Journal of the American Academy of Child and Adolescent Psychiatry* 36: 637–644.

Wassenaar, D. R. 2006. Commentary: Ethical considerations in international research collaboration: The Bucharest Early Intervention Project. *Infant Mental Health Journal* 27: 577–580.

Weaver, I. C. G., N. Cervoni, F. A. Champagne, A. C. D'Alessio, S. Sharma, J. R. Seckl, S. Dymov, M. Szyf, and M. Meaney. 2004. Epigenetic programming by maternal behavior. *Nature Neuroscience* 7: 847–854.

Wechsler, D. 1949. The Wechsler intelligence scale for children. New York: Psychological Corporation.

Wechsler, D. 1967. Manual for the Wechsler Preschool and Primary Scale of Intelligence. San Antonio, TX: Psychological Corporation.

Wechsler, D. 1989. Wechsler Preschool and Primary Scale of Intelligence —Revised. San Antonio, TX: Psychological Corporation.

Wechsler, D. 2003. WISC-IV technical and interpretive manual. San Antonio, TX: Psychological Corporation.

Wechsler, D. 2004. The Wechsler intelligence scale for children—fourth edition. London: Pearson Assessment.

Werker, J. F., and R. C. Tees. 2005. Speech perception as a window for understanding plasticity and commitment in language systems of the brain. *Developmental Psychobiology* 46 (3): 233–251.

Widom, C. S., K. DuMont, and S. J. Czaja. 2007. A prospective investigation of major depressive disorder and comorbidity in abused and neglected children grown up. *Archives of General Psychiatry* 64: 49–56.

Wiik, K., M. M. Loman, J. Van Ryzin, J. M. Armstrong, M. J. Essex, S. D. Pollak, and M. R. Gunnar. 2011. Behavioral and emotional symptoms of post-institutionalized children in middle childhood. *Journal of Child Psychology and Psychiatry* 52 (1): 56–63.

Williamson, J., and A. Greenberg. 2010. Families, not orphanages. Better Care Network Working Paper. Available at http://www.crin.org/docs/Families%20Not%20Orphanages.pdf.

Windsor, J., J. P. Benigno, C. A. Wing, P. J. Carroll, S. F. Koga, C. A. Nelson, C. H. Zeanah. 2011. Effect of foster care on young children's language learning. *Child Development* 82 (4): 1040–1046.

Windsor, J., L. E. Glaze, S. F. Koga, and the Bucharest Early Intervention Project Core Group. 2007. Language acquisition with limited input: Romanian institution and foster care. *Journal of Speech, Language, and Hearing Research* 50 (5): 1365–1381.

Windsor, J., A. Moraru, C. A. Nelson, N. A. Fox, and C. H. Zeanah. 2013. Effect of foster care on language learning at eight years: Findings from the Bucharest Early Intervention Project. *Journal of Child Language* 40: 605–627.

Yarrow, M. R., P. Scott, L. de Leeuw, and C. Heinig. 1962. Child-rearing in families of working and nonworking mothers. *Sociometry* 25 (2): 122–140.

Zamfir, C., ed., with contributions of E. Zamfir. 1998. *Toward a child-centered society: A report of the Institute for the Research of the Quality of Life.* Bucharest: Editura Alternative.

Zeanah, C. H., H. Egger, A. T. Smyke, C. Nelson, N. Fox, P. Marshall, and D. Guthrie. 2009. Institutional rearing and psychiatric disorders in Romanian preschool children. *American Journal of Psychiatry* 166: 777–785.

Zeanah, C. H., N. A. Fox, and C. A. Nelson. 2012. Case study in ethical issues in research: The Bucharest Early Intervention Project. *Journal of Nervous and Mental Disease* 200: 243–247.

Zeanah, C. H., M. R. Gunnar, R. B. McCall, J. M. Kreppner, and N. A. Fox. 2011. Children without permanent parents: Research, practice, and policy. VI: Sensitive periods. *Monographs of the Society for Research in Child Development* 76 (4, serial no. 301): 147–162.

Zeanah, C. H., S. K. Koga, B. Simion, A. Stanescu, C. Tabacaru, N. A. Fox, C. A. Nelson, and the Bucharest Early Intervention Project Core Group. 2006a. Ethical issues in international research collaboration: The Bucharest Early Intervention Project. *Infant Mental Health Journal* 27: 559–576.

Zeanah, C. H., S. K. Koga, B. Simion, A. Stanescu, C. Tabacaru, N. A. Fox, C. A. Nelson, and the Bucharest Early Intervention Project Core Group. 2006b. Ethical dimensions of the BEIP: Response to commentary. *Infant Mental Health Journal* 27: 581–583.

Zeanah, C. H., J. A. Larrieu, S. S. Heller, J. Valliere, S. Hinshaw-Fuselier, Y. Aoki, and M. Drilling. 2001. Evaluation of a preventive intervention for maltreated infants and toddlers in foster care. *Journal of the American Academy of Child and Adolescent Psychiatry* 40: 214–221.

Zeanah, C. H., C. A. Nelson, N. A. Fox, A. T. Smyke, P. Marshall, S. W. Parker, and S. Koga. 2003. Designing research to study the effects of institutionalization on brain and behavioral development: The Bucharest Early Intervention Project. *Development and Psychopathology* 15 (4): 885–907.

Zeanah, C. H., C. Shauffer, and M. Dozier. 2011. Foster care for young children: Why it must be developmentally informed. *Journal of the American Academy of Child and Adolescent Psychiatry* 50: 1199–2001.

Zeanah, C. H., and A. T. S. Smyke. 2005. Building attachment relationships following maltreatment and severe deprivation. In *Enhancing early attachments: Theory, research, intervention and policy,* edited by L. Berlin, Y. Ziv, L. Amaya-Jackson, and M. Greenberg, 195–216. New York: Guilford Press.

Zeanah, C. H., A. T. Smyke, S. Koga, E. Carlson, and the Bucharest Early Intervention Project Core Group. 2005. Attachment in institutionalized and community children in Romania. *Child Development* 76: 1015–1028.

Zeanah, C. H., A. T. Smyke, and L. Settles. 2006c. Children in orphanages. *Blackwell handbook of early childhood development,* edited by K. McCartney and D. Phillips, 224–254. Malden, MA: Blackwell.

Zelazo, P. D., U. Muller, D. Frye, S. Marcovitch, G. Argitis, J. Boseovski, J. K. Chiang, D. Hongwanishkul, B. V. Schuster, A. Sutherland, and S. M. Carlson. 2003. The development of executive function in early childhood. *Monographs of the Society for Research in Child Development* 68 (3).

Notes

1. The Beginning of a Journey

1. See, for example, Black et al., 1998; Greenough et al., 1987.

2. See Fox et al., 2010.

3. Rauscher et al., 1993.

4. For details about how the science of early development was often portrayed in the media, see Thompson and Nelson, 2001.

5. See Bruer, 1998.

6. Other members of the network included David Amaral, Ph.D., Judy Cameron, Ph.D., B.J. Casey, Ph.D., Allison Doupe, M.D., Ph.D., Eric Knudsen, Ph.D., Pat Levitt, Ph.D., Susan McConnell, Ph.D., Jack Shonkoff, Ph.D., and the late Marian Sigman, Ph.D.

7. Spitz, 1945.

8. See reports from the National Scientific Council on the Developing Child: http://developingchild.harvard.edu/resources/reports_and_working_papers/; also see Shonkoff and Phillips, 2000.

9. See, for example, Weaver et al., 2004.

10. See Francis and Meaney, 1999.

11. Weaver et al., 2004.

12. Szyf et al., 2007.

13. Rosapepe, 2001.

14. Rosenberg et al., 1992.

15. See Harlow, 1958; 1962.

16. Chisholm et al., 1995; Chisholm, 1998; Rutter and the English and Romanian Adoptees Study Team, 1998; O'Connor et al., 2000.

2. Study Design and Launch

1. Smyke et al., 2007.

3. The History of Child Institutionalization in Romania

1. Jarriel, 1990.

2. For more detail about the creation of child institutions and the pronatalist policies of the Ceauşescu period, see Kligman, 1998. For a review of the

reforms in child protection in the decade post-revolution, and an evaluation of their effects, see work by Greenwell, 2003; Popa-Mabe, 2010.

3. Zeanah et al., 2006c.

4. Viazzo et al., 2000.

5. Zeanah et al., 2006c.

6. Bachman, 1991.

7. Greenwell, 2003.

8. Crenson, 1998; Hacsi, 1997.

9. Boswell, 1988.

10. Rousseau, *Emile,* Book II.

11. Crenson, 1998; Smith and Fong, 2004.

12. Spitz, 1945. Goldfarb, 1943; 1945a; 1945b. Chapin, 1915.

13. Bowlby, 1951.

14. Zeanah et al., 2006c.

15. Child Welfare League of America, 2004.

16. Browne and Hamilton-Giachritsis, 2004.

17. Popa-Mabe, 2010.

18. Greenwell, 2003.

19. Popa-Mabe, 2010.

20. Popa-Mabe, 2010.

21. Kligman, 1998.

22. United Nations Development Program-Romania, 2002.

23. Kligman, 1998.

24. Greenwell, 2003.

25. Kligman, 1998; Greenwell, 2003.

26. Greenwell, 2003.

27. Greenwell, 2003.

28. Greenwell, 2003; Kligman, 1998.

29. Kligman, 1998.

30. Kligman, 1998.

31. Kligman, 1998.

32. Greenwall, 2003; Kligman, 1998.

33. Greenwell, 2003.

34. Greenwell, 2003.

35. Johnson et al., 1992.

36. Rosenberg et al., 1992.

37. Zamfir, 1998.

38. Tobis, 2000.

39. Skeels, 1966.

40. Stern, 1985.

41. Johnson et al., 1992.

42. Zaknun et al., 1991, as cited in Johnson et al., 1992.

43. Lie and Murarasu, 2001; Rosenberg et al., 1992.

44. Kline et al., 2007.

45. Novotny et al., 2003.

46. Dente and Hess, 2006.

47. Kligman, 1998.

48. Rosenberg et al., 1992.

49. Norris, 2009; Greenwell, 2003; Popa-Mabe, 2010.

50. Greenwell, 2003.

51. Zamfir, 1998.

52. Human Rights Watch, 1999, p. 420.

53. G. Koman, interview, January 2011.

54. Greenwell, 2003.

55. Brodzinsky and Schecter, 1990.

56. Hunt, 1991; Kligman, 1998.

57. Hunt, 1991.

58. Greenwell, 2003.

59. Greenwell, 2003.

60. Greenwell, 2003.

61. Greenwell, 2003.

62. See, for example, Kuddo, 1998, as cited in Tobis, 2000.

63. Tobis, 2000.

64. Greenwell, 2003.

65. Greenwell, 2003.

66. Tobis, 2000.

67. C. Tabacaru, interview, March 28, 2011.

68. Sandu, 2006.

69. Scheffel, 2005.

70. Roy, 2010.

71. Rosapepe, 2001.

72. Roy, 2010.

4. Ethical Considerations

1. We have deliberated and written about the ethical dimensions of this project from the outset. We thank the following journals for allowing us to develop our research in previous publications: *Science* 318 (supplementary online material); *Infant Mental Health Journal* 27: 559–576 and 581–583; and

Journal of Nervous and Mental Disease 200: 243–247. Information in this chapter also comes from commentaries by others: see Miller, 2009; Millum and Emanuel, 2007; *Nature Neuroscience,* 2008; Rid, 2012; and Wassenaar, 2006.

2. Zeanah et al., 2006c.

3. Rosapepe, 2001.

4. Zeanah et al., 2006c.

5. IRB Procedures, 1991, p. 6.

6. Children's Bureau, 2009.

7. See Zeanah et al., 2006b.

8. Ainsworth et al., 1978.

9. Emanuel et al., 2000; Miller, 2009; Millum and Emanuel, 2007; Rid, 2012.

10. Miller, 2009, quoting Beecher, 1966.

11. Miller, 2009; Rid, 2012.

12. Boothby et al., 2012.

13. See MacKenzie, 1999.

14. Miller and Brody, 2003; quote p. 25.

15. Child Welfare League of America, 2007.

16. National Authority for Child Protection and Adoption, 2004; Child Welfare League of America, 2007.

17. Council for International Organizations of Medical Sciences, 2002.

18. Nelson et al., 2007, supplementary online material.

19. Interview January 2011.

20. Groza et al., 2011.

21. Browne et al., 2006; Walker et al., 2011.

22. Millum and Emanuel, 2007.

23. United Nations, 1989.

24. Interview with Adina Codres, January 2011.

25. Interview with Bogdan Simion, January 2011.

26. Miller, 2009; Millum and Emanuel, 2007; Nature Neuroscience, 2008; Rid, 2012; Wassenaar, 2006. But see Loue, 2004.

5. Foster Care Intervention

1. See Larrieu and Zeanah, 1998.

2. We thank the following publications for allowing us to develop our research, in detail, in previous sources: *Science* 318 (supplementary online material); *Child and Adolescent Psychiatric Clinics of North America* 18: 721–734; *Enhancing early attachments: Theory, research, intervention and policy* (New York: Guilford Press), 195–216. Much of the information in this chapter comes from those accounts.

3. See Tobis, 2000; Browne, 2006; Walker et al., 2011.

4. See Tobis, 2000.

5. Dickens and Groza, 2004.

6. Tobis, 2000.

7. Simion, 2011.

8. Interview with Bogdan Simion, 2011.

9. Groza, 2001.

10. Marin Mic, personal communication, March 3, 2011.

11. Interview with Adina Codres, 2011.

12. Dickens and Groza, 2004.

13. Tobis, 2000.

14. See Dickens and Groza, 2004, for efforts in the 1990s to introduce concepts like individual empowerment and strengths-based practice into Romania.

15. Larrieu and Zeanah, 1998; Zeanah et al., 2001.

16. See Dozier et al., 2006; Zeanah et al., 2011.

17. National Research Council and Institute of Medicine, 2000; Zeanah and Smyke, 2005.

18. Heller et al., 2002; Smyke et al., 2009.

19. Dozier et al., 2006.

20. Gilkerson, 2004.

21. Smyke et al., 2009.

22. Interview with Cristian Tabacaru, April 2011.

23. See Craciun, 2011 (January 18, http://www.evz.ro/detalii/stiri/abandonati-si-de-parinti-si-de-mamele-sociale-918471.html, downloaded June 22, 2011).

6. Developmental Hazards of Institutionalization

1. Pinheiro, 2006.

2. Bowlby, 1944.

3. Goldfarb, 1943, 1944, 1945a, 1945b.

4. Freud and Burlingham, 1943.

5. Spitz, 1952.

6. Bowlby, 1951.

7. Skeels and Skodak, 1965. First described by Goldfarb (1943) and then confirmed by, among others, Provence and Lipton (1962) in the United States, and Tizard and Rees (1974) in Great Britain, children raised in institutions without the benefit of caregiver stimulation showed significant delays in intelligence.

8. http://rmc.library.cornell.edu/homeEc/cases/apartments.html.

9. Greenough et al., 1987.

10. For example,Yarrow et al., 1962; Rheingold, 1961.

11. Gleason et al., 2011.

12. Dennis and Najarian, 1957; Goldfarb, 1945a, 1945b; Provence and Lipton, 1962; Spitz, 1945.

13. Provence and Lipton, 1962.

14. Hunt et al., 1976; Provence and Lipton, 1962.

15. Levy, 1947; Tizard, 1977.

16. Dennis and Najarian, 1957.

17. Provence and Lipton, 1962.

18. Sonuga-Barke et al., 2008.

19. Sonuga-Barke et al., 2008.

20. van IJzendoorn et al., 2008.

21. van IJzendoorn et al., 2008.

22. Goldfarb, 1945a, 1945b; Spitz 1945; and later, Provence and Lipton, 1962.

23. Kaler and Freeman, 1994.

24. Dobrova-Krol, van IJzendoorn, et al., 2010.

25. Dobrova-Krol, van IJzendoorn, et al., 2010.

26. Sparling et al., 2005.

27. Tizard and Hodges, 1978; Tizard and Rees, 1974; Hodges and Tizard, 1989a.

28. Tizard and Rees, 1974.

29. Tizard and Rees, 1974.

30. Tizard and Hodges, 1978.

31. Hodges and Tizard, 1989a.

32. Goldfarb, 1945a, 1945b.

33. Vorria et al., 2003.

34. For example, Rutter et al., 1999; Ames and Carter, 1992; Gunnar and Kertes, 2003; and Pollak et al., 2010.

35. van den Dries et al., 2010.

36. van IJzendoor, Luijk, and Juffer, 2008.

37. Ames and Carter, 1992.

38. Rutter et al., 1998.

39. Rutter et al., 1998.

40. Beckett et al., 2006.

41. Tizard and Hodges, 1978.

42. Sonuga-Barke et al., 2008.

43. Pollak et al., 2010.

44. Colvert, Rutter, and Kreppner et al., 2008.

45. Dennis and Najarian, 1957; Goldfarb, 1945a; Provence and Lipton, 1962; Spitz, 1945.

46. Brodbeck and Irwin, 1946.

47. Brodbeck and Irwin, 1946.

48. Gindis, 2000.

49. Stromswold, 1995; Tarabulsy et al., 1996.

50. Hough, 2000.

51. Provence and Lipton, 1962.

52. Goldfarb, 1943; Lowrey, 1940.

53. Goldfarb, 1943; Provence and Lipton, 1962.

54. Tizard and Rees, 1974.

55. Vorria et al., 2003.

56. Juffer et al., 2005.

57. Marcovitch et al., 1997; Chisholm, 1998; O'Connor et al., 2003.

58. For example, Doolittle, 1995; Goldfarb, 1944; Roy et al., 2000; Wiik et al., 2011.

59. Goldfarb, 1943.

60. Doolittle, 1995; Kreppner et al., 2001; Roy et. al., 2000; Rutter et al., 2001.

61. Roy et al., 2000.

62. Goldfarb, 1944.

63. Lowrey, 1940; Tizard and Rees, 1974.

64. Rutter et al., 1999.

65. Chugani et al., 2001.

66. Eluvathingal et al., 2006

67. Chugani et al., 2001.

68. Tottenham and Sheridan, 2009; Tottenham et al., 2011.

69. Tottenham and Sheridan, 2009.

70. Tottenham et al., 2011.

71. Adolphs and Spezio, 2006.

72. Mehta et al., 2009.

7. Cognition and Language

1. Binet and Simon, 1904, 1916; Terman and Merrill, 1937.

2. Bayley, 1993.

3. Weschler, 1989; Wechsler, 2004.

4. Smyke et al., 2007.

5. NICHD Early Child Care Research Network, 1996.

6. Nelson et al., 2007.

7. Rutter, 1996.

8. Fox et al., 2010.

9. For example, Ramey and Campbell, 1984.

10. See, for example, Zelazo et al., 2003; Nelson and Luciana, 2008; Blair et al., 2005.

11. For example, Sanchez et al., 2001; Liston et al., 2009; Parker et al., 2005a.

12. Huttenlocher, 1979; Huttenlocher, 1984; Huttenlocher and de Courten, 1987; Huttenlocher and Dabholkar, 1997; Sowell et al., 1999; for a review, see Nelson and Jeste, 2008. For sensitivity to experience-dependent fine-tuning see, for example, Black et al., 1998.

13. Nelson, 1998; Champagne et al., 2008; Hill and McEwen, 2010.

14. Curtis et al., 2002.

15. Loman et al., 2013; Nelson, 2007; Nelson et al., 2007; Pollak et al., 2010.

16. Zelazo et al., 2003.

17. For example, Luciana and Nelson, 1989; Curtis et al., 2002; Luciana and Nelson, 2002.

18. Kochanska et al., 1998.

19. Cragg and Nation, 2008. See also Casey et al., 1997; Durston et al., 2002; Hwang et al., 2011.

20. Colvert, Rutter, Kreppner, et al., 2008; Pollak et al., 2010.

21. Beckett et al., 2007.

22. Loman et al., 2009.

23. See McDermott et al., 2012.

24. Bunge and Zelazo, 2006; Casey et al., 1997; Bunge and Wright, 2007.

25. Falkenstein et al., 1991.

26. Coles, 2001; Hermann et al., 2005; Roger et al., 2010; van Veen and Carter, 2002.

27. Loman et al., 2013.

28. McDermott et al., 2013.

29. See Luciana and Nelson, 1998; Luciana and Nelson, 2002.

30. See Bos et al., 2009.

31. Hart and Risley, 1995; Noble et al., 2007.

32. See Windsor et al., 2007, 2011.

33. Windsor et al., 2007.

34. Windsor et al., 2011.

35. Werker and Tees, 2005.

8. Early Institutionalization and Brain Development

1. For a tutorial on the use of EEG in developmental studies, see de Haan, 2007.

2. Marshall et al., 2002.

3. Barry et al., 2003.

4. Marshall et al., 2004.

5. Marshall et al., 2008.

6. Vanderwert et al., 2010.

7. For a review of the ERP literature in the context of development, see Nelson and Monk, 2001; Nelson and McCleery, 2008.

8. For a review, see de Haan et al., 2002.

9. As reviewed in Zeanah et al., 2003.

10. Parker et al., 2005a; for ERP at later ages see Moulson et al., 2009. Moulson, Westerlund, et al., 2009.

11. Righi and Nelson, 2013.

12. Nelson, 2001.

13. For example, Fries et al., 2004; Pollak, 2005.

14. See Parker et al., 2005b.

15. Moulson, Fox, et al., 2009; Moulson, Westerlund et al., 2009.

16. For example, de Haan and Nelson, 1997.

17. Nelson et al., 2006; Jeon et al., 2010.

18. For details, see Nelson et al., 2013.

19. Generally in adolescence; see Thomas et al., 2007.

20. Moulson, Fox, et al., 2009, discussed above.

21. See Nelson and de Haan, 1997.

22. See Sheridan et al., 2012.

23. For enlarged amygdala, see Tottenham et al., 2010, and Mehta et al., 2009. For no change, see Gunnar and Fisher, 2006.

24. Huttenlocher, 2002.

9. Growth, Motor, and Cellular Findings

1. Johnson et al., 1992; Albers et al., 1997; Miller and Hendrie, 2000.

2. Powell et al., 1967.

3. Johnson, 2011.

4. See Johnson et al., 2010, for details.

5. Sheridan et al., 2012.

6. This has been reported by Cermak and Daunhauer, 1997; and Harris et al., 2008.

7. Fisher et al., 1997; Smyke et al., 2002.

8. Bos et al., 2010.

9. Bos et al., 2010.

10. Loman et al., 2009.

11. Second ed., short form: Bruininks and Bruininks, 2005.

12. Levin et al., 2013.

13. McEwen, 1998; Phillips et al., 2005.

14. Gilley et al., 2008.

15. Fitzpatrick et al., 2007; Martin-Ruiz et al., 2006.

16. Epel et al., 2004; Kananen et al., 2010; Simon et al., 2006; Damjanovic et al., 2007; Lung et al., 2007; Parks et al., 2009.

17. Drury et al., 2011.

18. Barker et al., 1989a, 1989b.

10. Socioemotional Development

1. See Zeanah et al., 2006c, and Chapter 6.

2. See Robertson and Robertson, 1989.

3. For social abnormalities, see Chisholm et al., 1995; Chisholm, 1998; Marcovitch et al., 1997; O'Connor et al., 1999. For emotional abnormalities, see Sloutsky, 1997; Fries and Pollak, 2004.

4. For details, see Ghera et al., 2009; Smyke et al., 2007.

5. Goldsmith and Rothbart, 1999.

6. Ghera et al., 2008.

7. Harlow and Suomi, 1970.

8. Bowlby, 1969, 1973, 1980.

9. Ainsworth et al., 1978.

10. Ainsworth et al., 1978.

11. See van IJzendoorn et al., 1992.

12. Main and Solomon, 1990.

13. Carlson et al., 1989.

14. Cyr et al., 2010; van IJzendoorn et al., 1999.

15. Cassidy et al., 1992.

16. Green and Goldwyn, 2002.

17. DeKlyen and Greenberg, 2008.

18. De Wolff and Van IJzendoorn, 1997.

19. DeKlyen and Greenberg, 2008; Green and Goldwyn, 2002; Zeanah et al., 2006c.

20. Zeanah et al., 2005.

21. Vorria et al., 2003.

22. Dobrova-Krol, Bakermans-Kranenburg, et al., 2010.

23. Tizard and Rees, 1975.

24. Zeanah et al., 2005.

25. Zeanah et al., 2005.

26. Zeanah et al., 2005.

27. Dobrova-Krol, Bakermans-Kranenburg, et al., 2010.

28. Gleason et al., 2011.

29. Tizard and Rees, 1975.

30. Tizard and Rees, 1975.

31. Hodges and Tizard, 1989b; Rutter et al., 2007; Rutter et al., 2009; Tizard and Hodges, 1978.

32. Zeanah et al., 2005.

33. Smyke et al., 2010.

34. Smyke et al., 2012.

35. Smyke et al., 2012.

36. Chisholm, 1998; Hodges and Tizard, 1989b; Rutter et al., 2007.

37. Gleason et al., forthcoming.

38. Smyke et al., 2012.

39. Smyke et al., 2012.

40. Roy et al., 2004.

41. Kaler and Freeman, 1994.

42. Erol et al., 2010.

43. Erol et al., 2010.

44. Erol et al., 2010.

45. See Parker et al., 2006, for a review.

46. Gresham and Elliot, 1990.

47. Almas et al., 2012.

48. Almas et al., in prep.

49. Kashy and Kenny, 1999; Kenny, 1996.

50. Almas et al., 2012.

11. Psychopathology

1. Netter and Magee, 2010.

2. Netter and Magee, 2010.

3. http://adoption.about.com/od/adoptionrights/p/russian_children_murdered_by_adoptive_parent.htm.

4. Bowlby, 1969, pp. xiii–xiv.

5. Rutter et al., 2010.

6. For attachment disorders see Chisholm et al., 1995, and Rosenberg

et al., 1992; for stereotypies see Cermak et al., 1997, and Marcovitch et al., 1997; for internalizing and externalizing behavior problems, see Fisher et al., 1997, and Marcovitch et al., 1997.

7. For example, Hoeksbergen et al., 2003; O'Connor et al., 1999; Roy et al., 2004.

8. Fisher et al., 1997; Marcovitch et al. 1997.

9. Carter et al., 2003.

10. Carter et al., 2003.

11. Smyke et al., 2007.

12. Egger et al., 1999, version 1.3.

13. Egger et al., 2006.

14. Gleason et al., 2011.

15. Fisher et al., 1999.

16. Kraemer et al., 2002.

17. Egger et al., 2006.

18. Volkmar et al., 2007.

19. Zeanah et al., 2009.

20. The role of genes and environments in psychopathology has been summarized by Moffitt et al., 2006.

21. Moffitt et al., 2006.

22. Gogos et al., 1998; Karoum, Chrapusta, and Egan, 1994.

23. See Drury et al., 2010.

24. Drury et al., 2010.

25. Belsky et al., 2007; Belsky and Pluess, 2009.

26. Drury et al., 2012.

27. Drury et al., 2012.

28. Kraemer et al., 2002.

29. For internalizing psychopathology: Allen et al., 1998; Lee and Hankin, 2009; Lyons-Ruth et al., 1997; McCartney et al., 2004; for anxiety disorders and major depression: Abela et al., 2005; Warren et al., 1997.

30. Zeanah et al., 2009.

31. Smyke et al., 2010.

32. McLaughlin et al., 2012.

33. McGoron et al., 2012.

34. Kreppner et al., 2001; Rutter et al., 2010; Stevens et al., 2008; Zeanah et al., 2009.

35. Moffitt et al., 2006.

36. McLaughlin et al., 2010.

37. Casey et al., 1997; Konrad et al., 2006; Castellanos et al., 1996; Mostofsky et al., 2002; Rubia et al., 1999; Halperin and Schultz, 2006.

38. Narr et al., 2009; Shaw et al., 2006.

39. Narr et al., 2009; Shaw et al., 2006; Shaw and Sharp, 2007.

40. McLaughlin et al., 2010.

41. Marshall et al., 2004; see also Chapter 8.

42. McLaughlin et al., 2010.

43. Davidson and Fox, 1982.

44. See McLaughlin et al., 2012.

45. McLaughlin et al., 2011.

46. Davidson, 2000.

47. Banks et al., 2007.

48. Kochanska et al., 2000; Miller, 2000.

49. For behavioral responses see Parker and Nelson, 2004; for ERPs see Moulson, Fox, et al., 2009, and Moulson, Westerlund, et al., 2009.

50. Pollak and Sinha, 2002.

51. Slopen et al., 2012.

52. McLaughlin et al., 2011.

12. Putting the Pieces Together

1. Greenough et al., 1987.

2. Le Grand et al., 2001.

3. William Greenough, personal communication.

4. Fox et al., 2011; Smyke et al., 2012.

5. Rutter et al., 2004.

6. Sheridan et al., 2012; Vanderwert et al., 2010.

7. See St. Petersburg–USA Orphanage Research Team, 2008. Berument, 2013.

8. Knudsen, 2004.

9. See Zeanah et al., 2011.

10. Maurer et al., 2006; Vanderwert et al., 2010.

11. Vanderwert et al., 2010.

12. Rutter et al., 2010.

13. Brestan and Eyberg, 1998.

14. Williamson and Greenberg, 2010; Save the Children, 2009.

15. Ahmad et al., 2005; Dennis and Najarian, 1957; Goldfarb, 1943, 1944, 1945a, 1945b; Harden et al., 2002; Levy, 1947; Provence and Lipton, 1962; Roy et al., 2000.

16. Zeanah, Smyke, and Settles, 2006c.

17. Crenson, 1998; Smith and Fong, 2004.

18. Crenson, 1998; Hacsi, 1997.

19. Mitroi, 2012.

20. Tizard and Rees, 1975; Rutter et al., 2009.
21. Zeanah et al., 2011.
22. Knudsen et al., 2006.
23. Fox et al., 2011.
24. Gelles, 2011; Smyke and Breidenstine, 2009; Zeanah et al., 2011.
25. Chou and Browne, 2008; The Hague Convention, 1993.
26. Cristian Tabacaru, interview, March 28, 2011.
27. Gabriella Koman, interview, January 2011.
28. Bogdan Simion, interview, January 2011.
29. See Groza et al., 2011.
30. PL 109–95 Fourth Annual Report to Congress, 2010.
31. Boothby et al., 2012.
32. Boothby et al., 2012.
33. Boothby et al., 2012.
34. Browne et al., 2004.

Acknowledgments

An undertaking as complex as the Bucharest Early Intervention Project (BEIP) could not have happened without much good fortune, considerable hard work, and significant contributions from many people and organizations.

We are deeply grateful to the John D. and Catherine T. MacArthur Foundation for supporting the research network that created and sustained BEIP and for support throughout the initial phase of the project. The flexibility afforded by this support allowed us to adapt and adjust in ways that were crucial to our success. Subsequently, the foundation provided start-up support for the Institute of Child Development in Bucharest, which has trained professionals who provide services to children throughout Romania. Other organizations and individuals also generously supported this project over the years: the James S. McDonnell Foundation, the Irving Harris Foundation, USAID, the Binder Family Foundation, Help the Children of Romania, Inc., the National Institutes of Mental Health (MH091363), the Center on the Developing Child at Harvard University, the Sinneave Family Foundation, Richard and Mary Scott, and several anonymous donors who have graciously contributed funds to the project.

We want to thank the families and children who have continued to participate in the BEIP since its inception. In a setting with little tradition for the kind of longitudinal, child development–based research that BEIP represents, it is amazing that they have stayed with us for so many years. We owe them our utmost gratitude.

Dana Johnson provided the original inspiration for this project. He appreciated the importance of bringing rigorous research to bear on crucial questions of policy about the lives of institutionalized children. Dana secured us a place on a medical mission trip to Romania in 1998, organized by Ronald Federici, a neuropsychologist with extensive experience in evaluating children who have spent time in institutions. With a pediatric nurse practitioner, Mary Jo Spencer, Dana originally screened all the potential participants in the study. He has remained involved as a valued research collaborator.

Many professionals in Romania have supported our project over the years; three in particular have been instrumental in its success. Cristian Tabacaru, the first secretary of state for child protection in Romania and later director of SERA Romania, invited us to conduct the project in 1999, and he provided essential support, advice, and help over the years. Bogdan Simion, who currently directs SERA Romania, oversaw construction of the original and new BEIP laboratories and has been a valued and trusted colleague and essential collaborator in every aspect of the project since its inception. These two colleagues continue to have a significant impact on our project and its success. Finally, Sebastian Koga took a leave from medical school and served as project director in Bucharest from 2000 to 2004 before returning to complete medical school and later a residency in neurosurgery. He was the first project coordinator, and his careful attention to detail was matched only by his grasp of the big picture. This was essential to the launch and early success of the project. From repairing the EEG equipment to hiring the staff to meeting with the American ambassador to Romania, Sebastian demonstrated a remarkable talent for navigating through challenges large and small.

Other Romanian colleagues who were instrumental in launching and sustaining BEIP include Mihai Iordachescu, who served as project pediatrician and later as a member of the data safety monitoring board; Alin Stanescu, a pediatrician who led the Institute of Maternal and Child Health in Bucharest; and Gabriela Coman, who assisted us from within the National Authority for Child Protection and later served as advocate and then as a member of the data safety monitoring board. Danut Fleaca, director of Sector 1 in Bucharest, generously supplied us with information on and access to the lives of children from his sector when needed. Adina Codres was director of St. Catherine's Placement Center when our project began and provided us with access and help. Anca Nicolae graciously assisted with early data collection and provided training to the staff when the project began. Additionally, other SERA Romania staff have been helpful over the years: Agustina Rouchdi and Stefan Pascu (human resources), Cristian Nistor (work safety), and Emilia Popa and Dana Sirbu (accounting). Finally, we are particularly grateful to three valuable colleagues at Bucharest University, Mihai Anitei, Nicolae Mitrophan, and Mircea Dumitru.

Our Bucharest staff is second to none. Anca Radalescu (2000–present) was working as a psychologist at St. Catherine's Placement Center, our future home base, when we met her. She became our first research assistant and has been indispensable ever since. Her knowledge of the children and families in the project and the child-protection system in Bucharest has proven invaluable, as has her consistent positive energy. Florin Tibu oversaw the fifty-four month assessment (2003–2007), and after a leave to obtain his Ph.D in Psychology, returned to BEIP for the twelve-year assessment (2010–present). We have benefited from his deep understanding of our goals, his skill in helping us achieve them, and his willingness to assist with any tasks. Nicoleta Corlan (2001–present) and Nadia Radu (2002–present) have been steadfast partners almost since the beginning, and both have made invaluable contributions to many aspects of the project. Alexandra Cercel (2010–present) joined the project more recently, but soon thereafter became essential to its success, mastering new procedures efficiently and effectively. Iuliana Dobre (2001–present) has served many key roles, beginning as a project assistant and now seving as assistant accountant, all with cheerfulness and dedication. We are also indebted to her for making sure we were well fed after our long trips from the United States to Bucharest and for providing a home away from home to several families who traveled from outside Bucharest to participate.

The original social work team of Alina Rosu (2001–2004), Veronica Ivascanu (2001–2005 and 2010–2012), and Amelia Greceanu (2001–2008) performed heroically in recruiting, training, and supporting a network of foster families when there was no tradition of that in Bucharest. More recently, Carmen Iuga (2008–present) and Mariana Mitu (2013–present) have maintained these high standards by continuing to provide support effectively and to advocate on behalf of the families in the project.

We also appreciate other staff who made key contributions along the way that helped make this project possible. Research assistants included Carmen Calancea (2000–2001), Roxana Nedelcu (2001–

2003), Adela Apetroaia (2001–2004), and Adina Ungureanu (2002–2006). Raluca Mandea was assistant project manager (2001), as was Magda Vlad (2002–2005), and Calin Gligorea (2007–2008) was later deputy project manager. Monica Ciobanu (2001–2005), Georgiana Carstea (2004–2008), and Emilia Popa (2003–present) provided accounting services. Dana Grozav (2003–2006) and Catalin Calusaru (2003–2004) served as data coders, as did the psychology graduate students from Bucharest University Maria Magdalena Stan, Anca Matei, Daniella Casella, and Diana Cramer Cosmescu. We also thank drivers who made hundreds of trips transporting children and families to the lab for assessments: Gabriel Calin (2001–2002), Florin Dobre (2002–2003), Matache Laurentiu (2003–2004), Iulian Petrea (2004–2008), and Nicolae Radoi (2008–2010).

Hermi Woodward and Marcy Ray served as MacArthur Foundation network administrators from 1997 to 2005 and played invaluable roles in the early days of developing and administering this project. Elizabeth Furtado (2005–present) continues to play an essential and much appreciated role as project coordinator, a job that requires remarkable skills of organization, diplomacy, patience, vision, and a high tolerance for the eccentricities of the three principal investigators. Elizabeth, we owe you our eternal gratitude.

We also have been fortunate to benefit from the collaboration of a number of U.S. investigators: Susan Parker (Randolph Macon University); Jennifer Windsor (University of Minnesota); Margaret Moulson (Ryerson University); Alissa Westerlund (Boston Children's Hospital); Karen Bos (San Mateo County Psychiatry); April Levin (Boston Children's Hospital and Harvard Medical School); Kate McLaughlin (University of Washington); Margaret Sheridan (Boston Children's Hospital and Harvard Medical School); Anna

Smyke, Mary Margaret Gleason, Stacy Drury, Lucy McGoron, Zoe Brett, and Kate Humphreys (Tulane University); Peter Marshall (Temple University); Alisa Almas (University of British Columbia); Ross Vanderwert (Boston Children's Hospital and Harvard Medical School); Kate Degnan (University of Maryland); Jennifer McDermott (University of Massachusetts); Connie Lamm (University of New Orleans); Sonya Troller-Renfree, Meg Woodbury, and Carl Lejuez (University of Maryland); Catalina Kopetz (Wayne State University); Helen Egger (Duke University); and Bethany Reeb Sutherland (Florida International University). We also wish to thank the undergraduates who have helped with this project, including Hana Jeon, Audrey Young, Winnie Lin, and Nora Kovar (Harvard College).

Julie Larrieu, Tammy Coots, and Valerie Wajda Johnston provided essential training and supervision in the early phases of the project. Joanna Santamaria conducted staff training on the WISC, and Devi Miron (Tulane University) consulted in Bucharest about WISC assessments at later ages. Roxanna Malene (University of Minnesota) and Lavinia Mitroi (Harvard College) played an invaluable role in translating documents for the BEIP team and coding several of our measures. We are also indebted to Michael Scheeringa for developing diagnostic algorithms and to Prudence Fisher for helping us with the DISC. Finally, Katherine Oberwager performed yeoman's work pulling together references and figures for the book.

Gwen Gordon, and later Jisun Jang, Kimberly Chin, and Lily Chen served as data managers of a complex and ever-enlarging data set. Don Guthrie was BEIP statistician extraordinaire until he retired; later, Matt Gregas, and now Kate Degnan and Caterina Stamoulis, provided valuable statistical input. Adina Chirita was a collaborator on obtaining MRIs in Bucharest.

All of us have benefited from discussions with dozens of colleagues from around the world, many of whom are well versed in the issues we describe in this book. There is no way to list them individually, but they have stimulated, challenged, and inspired us in various ways.

United States Senator Mary Landrieu and Kathleen Strottman, executive director of the Congressional Coalition for Adoption Institute, were extraordinarily helpful to us soon after the project began and have remained supportive. United States ambassador to Romania, Michael Guest, and later Ambassador Mark Gitenstein also supported our efforts in important ways.

Finally, we are thankful to the Rockefeller Foundation for the month the three of us spent in the residency program of their Bellagio Center in summer 2011. Without that protected time, it would have been far more challenging for us to complete this endeavor. The stimulating and supportive environment of the center allowed us to complete an initial draft, as well as to engage in lively discussions about the material with one another and with other residents at the center. Lake Como and the surrounding Italian villages are forever with us, and the support we received there essential.

This book is about the power of relationships. Through it, our relationships with one another and with many others have been immeasurably enriched, more than we could have imagined.

Index

Note: Page numbers followed by an f or t indicate figure and table respectively.